SCIENCE FAIR WARM-UP
» LEARNING THE PRACTICE OF SCIENTISTS «
Teachers Guide

SCIENCE FAIR WARM-UP

» LEARNING THE PRACTICE OF SCIENTISTS «

Teachers Guide

JOHN HAYSOM

National Science Teachers Association

Arlington, Virginia

National Science Teachers Association

Claire Reinburg, Director
Jennifer Horak, Managing Editor
Andrew Cooke, Senior Editor
Wendy Rubin, Associate Editor
Agnes Bannigan, Associate Editor
Amy America, Book Acquisitions Coordinator

ART AND DESIGN
Will Thomas Jr., Director
Rashad Muhammad, Graphic Designer
Illustrations by Bob Seguin

PRINTING AND PRODUCTION
Catherine Lorrain, Director
Jack Parker, Electronic Prepress Technician

NATIONAL SCIENCE TEACHERS ASSOCIATION
Gerald F. Wheeler, Executive Director
David Beacom, Publisher

1840 Wilson Blvd., Arlington, VA 22201
www.nsta.org/store
For customer service inquiries, please call 800-277-5300.

FSC
www.fsc.org
MIX
Paper from
responsible sources
FSC® C011935

NSTA is committed to publishing material that promotes the best in inquiry-based science education. However, conditions of actual use may vary, and the safety procedures and practices described in this book are intended to serve only as a guide. Additional precautionary measures may be required. NSTA and the authors do not warrant or represent that the procedures and practices in this book meet any safety code or standard of federal, state, or local regulations. NSTA and the authors disclaim any liability for personal injury or damage to property arising out of or relating to the use of this book, including any of the recommendations, instructions, or materials contained therein.

Cataloging-in-Publication Data is available from the Library of Congress for each student edition.

ISBN 978-1-936959-23-5

eISBN 978-1-936959-68-6

Contents

Contents

Note to Teachers

There are three student books in the *Science Fair Warm-Up* series:

- The book for grades 5–8 includes all the one-star lessons. (See Content Overview, page xxii.)

- The book for grades 7–10 includes the two-star lessons.

- The book for grades 8–12 includes the three-star lessons.

Teachers may wonder which student book or books best meet their needs. Some might be puzzled about how it is possible for a book to be appropriate for such a wide range of grades.

The book for grades 5–8 is particularly suitable for students who have not participated in a science fair before and lays a foundation for the ideas developed in the later books about the practices of scientists. Indeed, even those students who have experienced science fairs before will undoubtedly encounter many ideas about scientific practices that are new to them.

The book for grades 7–10 develops the ideas about practices that students encounter in the first book. Students will also find that the two-star problems are much more cognitively demanding.

The book for grades 8–12 further develops the ideas about the practices of scientists. In addition, many of the problems the students will encounter are challenging, so much so that in field testing the book has been used with both grade 8 students and university science graduates.

It is anticipated that middle school teachers will find the series valuable when they use the first book in grade 6, the second book in grade 7, and the third book in grade 8. In addition, the series provides high school teachers with a curriculum that uniquely meets many of the goals outlined in *A Framework for K–12 Science Education: Practices, Crosscutting Concepts, and Core Ideas* (NRC 2012).

ACKNOWLEDGMENTS

The Story of a Curriculum Project

The stimulus to write this book has come from many people. In the first place, I was motivated by the students themselves. I met many students while judging at science fairs. I was frequently struck by their enormous enthusiasm and pride in what they had done. I also enjoyed the experience of helping my own three children with their projects. I watched their struggles and struggled myself as I tried to assist them—without providing too many solutions to the problems they encountered or taking ownership of the projects from them. The rich educational potential of students undertaking scientific investigations—coupled with an apparent lack of suitable curriculum materials—prompted me to begin this work, which spanned some six or seven years.

I began by asking the students (and their teachers) at Ellenvale School in Dartmouth, Nova Scotia, to share their project work with me. Over the years, I viewed their displays and read hundreds of their reports. This process provided a large bank of interesting ideas for activities and projects and an in-depth appreciation of some of the frequently occurring problems and difficulties students and teachers encountered. I am grateful to the students and teachers for sharing their experiences with me and have included many extracts from their projects.

The curriculum design process that followed was not a linear one, and it evolved slowly. In retrospect, it appears to be somewhat similar to piecing together a three-dimensional jigsaw in which many of the pieces were missing! The first dimension involved relating the problems and difficulties that students met (both at Ellenvale School and elsewhere) to our understanding of how scientists actually work in practice, an understanding that is distinctly different from the common but mistaken view that there is such a thing as a scientific method. The second dimension involved perceiving the problems and difficulties as opportunities for learning and, with this in mind, searching for and devising appropriate learning experiences. Finally, it became evident that some of the learning experiences were more sophisticated than others, either requiring some previous understanding of scientific inquiry or being more intellectually demanding.

The first draft of the materials was used by my student-teachers at Saint Mary's University who were helping teenagers with their project work in a laboratory-school setting. The student-teachers' feedback was frank and formative, and I would like to acknowledge its importance.

I used an expanded second draft of the materials in a local school and was grateful to be able to share my experience with a number of other

Acknowledgments

teachers who had kindly offered to field-test the materials in their classrooms. I would particularly like to acknowledge the continued help I have received from Karen Getson and Bob Dawson at Ellenvale.

It would not have been possible to field-test the materials at all without the assistance of Sue Conrad at Saint Mary's. She devoted many hours to designing and illustrating the student text. Her influence on its style is apparent.

The final draft was field-tested by still more teachers and was critically but constructively reviewed by Earl Morrison of the Atlantic Science Curriculum (ASCP) Project. Earl and my friends on the Board of ASCP have been a continual source of encouragement. I sincerely appreciate the help that all these people and other colleagues have given me over the years.

In conclusion, I would like to thank Professor Derek Hodson, an authority on the philosophy and sociology of science, for his helpful and insightful comments on the manuscript.

—*John Haysom*

About the Author

After completing his PhD in chemistry at Cambridge University, John Haysom taught science in a variety of schools before becoming a member of the faculties of education at five universities: Oxford University, Reading University (UK), University of the West Indies, Mount Saint Vincent University (Canada) and Saint Mary's University (Canada), where he is Professor Emeritus.

John has gained an international reputation as a teacher-educator and curriculum developer. In the UK, he was the coordinator of the groundbreaking Science Teacher Education Project, funded by the Nuffield Foundation. This was probably the first teacher education curriculum project in the world and was adapted for use in Australia, Canada, Israel, and other countries. At the University of the West Indies, he was responsible for the design and implementation of an innovative, theme-based inservice B.Ed. curriculum. As professor of education at Saint Mary's University, he initiated and helped lead the Atlantic Science Curriculum Project's SciencePlus textbook series. This curriculum was highly rated and became widely adopted in the United States. He has acted as a science curriculum consultant to the government of Trinidad and Tobago and to a number of projects in the United States.

John is the author of many books for teacher educators, teachers, and schoolchildren, as well as academic papers in curriculum design, evaluation and implementation, and teacher education. His most recent work includes *Predict, Observe, Explain: Activities Enhancing Scientific Understanding* (with Michael Bowen; NSTA 2010) and "Science Curriculum Research and Development in Atlantic Canada: A Retrospective," an article prepared for the *Canadian Journal of Science, Mathematics and Technology Education* (in press).

John has long been interested in science fairs, both as a teacher and a judge, and was elected to the board of directors of the Youth Science Foundation, the national body responsible for science fairs in Canada.

Introduction

Opportunities—wonderful opportunities—are provided in middle schools and high schools across North America for students to participate in science fairs or carry out their own science projects. Although many teachers believe it's valuable for their students to experience genuine scientific inquiry, many teachers also find the challenge a difficult one to meet. Some teachers simply resort to assigning a project for students to complete. Others spend hours doing their best to help more than 100 students individually. Many teachers dread the annual science fair. These curriculum materials are designed to help teachers create courses and programs that address their problems and needs. Two questions loom large in the minds of many teachers:

1. How can you organize 30 students doing 30 different projects at the same time?

2. How can you assist students and, at the same time, provide them with the freedom of choice and independence of thought that characterize genuine inquiry?

These materials address these problems. Because science fair projects frequently involve experimental work, the materials in this book focus on experimental work as well.

It certainly is valuable for students to experience genuine scientific inquiry. The experience provides them with the opportunity to become problem solvers (scientific problem solvers) and gain an understanding of the art of solving problems (the nature of scientific inquiry).

Being a good problem solver is quite different from understanding the problem-solving process. It is one thing to be able to find out for yourself which type of flashlight battery is better and another thing to appreciate or critique the way in which somebody else attacked the problem.

Although these ends are valuable in and of themselves, they also provide a valuable perspective on the way scientists have discovered the truly powerful network of explanations and ideas that enables us to make meaning of our natural world. For many people, it's a mystery where these explanations and ideas came from. For them, the connection between the products of science and people in white coats doing experiments is a tenuous one.

Most of today's science curricula focuses appropriately on helping students understand this network of explanations and ideas, but many students find this a difficult and meaningless task, one devoid of humanity and reality. We might be able to help these students by presenting science as a product of human endeavor. These curriculum materials attempt to do just this and thus might be used to enhance and enrich a curriculum with a product orientation.

Both the *National Science Education Standards* (NRC 1996) and the recently published *A Framework for K–12 Science Education* (NRC 2012) acknowledge these important goals. The Standards has this to say:

> The standards on inquiry highlight the ability to conduct inquiry and develop understanding about scientific inquiry. Students at all grade levels and in every domain of science should have the opportunity to use scientific inquiry and develop the ability to think and act in ways associated with inquiry, including asking questions, planning and conducting investigations, using appropriate tools and techniques to gather data, thinking critically and logically about relationships between evidence and explanations, constructing and analyzing alternative explanations, and communicating scientific arguments. (NRC 1996, p. 105)

A Framework for K–12 Science Education (NRC 2012) is a remarkable document. It is divided into three major sections: practices, crosscutting concepts, and core ideas. The word *practices* is closely related to ways in which the phrase *scientific inquiry* is used in this book. Incidentally, teachers who wish to deepen their understanding of the nature of science and its importance for science education will undoubtedly find the Framework (and some of the references it contains) most illuminating. Here are a few extracts from the beginning of the section that discuss what is meant by *practices*.

> Understanding How Scientists Work. The idea of science as a set of practices has emerged from the work of historians, philosophers, psychologists, and sociologists over the past 60 years.

This work illuminates how science is actually done … (NRC 2012, p. 43)

Our view is that this perspective is an improvement over previous approaches in several ways. First, it minimizes the tendency to reduce scientific practice to a single set of procedures, such as identifying and controlling variables, classifying entities, and identifying sources of error. This tendency overemphasizes experimental investigation at the expense of other practices, such as modeling, critique, and communication. (NRC 2012, p. 43)

Second, a focus on practices (in the plural) avoids the mistaken impression that there is one distinctive approach common to all science—a single "scientific" method … (NRC 2012, p. 44)

Third, attempts to develop the idea that science should be taught through a process of inquiry have been hampered by a lack of a commonly accepted definition of its constituent elements. (NRC 2012, p. 44)

These curriculum materials share these views and are in harmony with the above perspective. Indeed, the materials' design was driven in part by the apparent lack of materials that present an authentic view of the nature of science and scientific inquiry.

The situation in Canada is similar. The Council of Ministers of Education's vision of scientific literacy is built on four foundations, the second of which is skills (CMEC 1997):

> Skills—Students will develop the skills required for scientific and technological inquiry, for solving problems, for communicating scientific ideas and results, for working collaboratively, and for making informed decisions.

These skills are grouped into four main areas: initiating and planning, performing and recording, analyzing and interpreting, and communication and teamwork. The scope and complexity of these skills are developed in grades K–12.

The sections that follow include a discussion of the views taken on the nature of science and scientific inquiry, the teaching strategy used, a consideration of the problem of helping the students develop their independence, an outline of the ways in which the materials have been structured, and an overview of the content.

Nature of Science and Scientific Inquiry

Every set of science curriculum materials inevitably takes a position on the nature of science and scientific inquiry. Indeed, the shape of science curriculum materials is significantly shaped by the designer's views of the nature of science and scientific inquiry. Seldom, however, do you find these views made explicit in teachers guides. At first sight, this may seem rather strange, but it is not an easy task for designers to make explicit their views, and moreover, it is contentious because of the divergence that exists.

In the five paragraphs that follow, I attempt to outline the position adopted in these materials.

a. Scientific inquiry is artful and intuitive.

Some curriculum materials present scientific inquiry as a linear process, one that follows a sequence such as observation, formulation of hypothesis, experiment design, collection of results, and development of conclusions. But do scientists actually proceed this way? How would scientists proceed to investigate bouncing balls? It's likely that many thoughts would go through

their heads: "The substance the ball is made of and the substance it strikes might make a difference. How can the height of bounce be accurately measured? What is known about elasticity already?" Why is it that there is a limit to how high a ball bounces regardless of the height from which it is dropped? Sometimes an experiment doesn't work out well. Sometimes a chance observation opens up new avenues of exploration. Scientific inquiry appears to be a messy business; good scientists are guided intuitively as to what to do next. Nobel Laureate Peter Medawar considered scientific reasoning to be a constant interplay of interaction between hypothesis and the logical expectations they give rise to, a restless to and fro motion of thought, a kind of dialogue between the possible and the actual (Medawar 1969). It is just as appropriate to start at the end as it is at the beginning, and although the materials are arranged in a fairly traditional sequence (e.g., from experiment to interpretation of data to explanation), this is not designed to be prescriptive.

The popular idea of the scientific method seriously distorts reality. It portrays the process of discovery as an algorithm that, if simply plugged into a problem, will inevitably unlock our entry to the secrets of the universe.

b. The thinking processes involved in scientific inquiry are interwoven.

Some curriculum materials emphasize the learning of science process skills (observing, classifying, predicting, measuring, inferring, and so on) and proceed to teach these independently of context. But what is the point of observing a candle, classifying buttons, or measuring the width of a table? Developing such skills is important, but they should be learned in context. Learning should be holistic. To do otherwise not only makes learning less meaningful for the students but also gives them a false impression of the nature of science and scientific inquiry.

In these curriculum materials, such skills are developed in context. Bean seed germination is observed with a view to understanding what might speed it up or slow it down, and measurements of magnetic strength are made with a view to finding out the best way to determine the strength of homemade magnets. Different lessons highlight different aspects of the process of inquiry but always put these in context.

c. The process, product, and purpose of scientific inquiry are intimately related.

Some curriculum materials distort the process of inquiry by separating it from the product. Moreover, an appreciation of both process and product remains limited unless we understand their purpose in terms of human interest. For example, when the City of London was being rebuilt after the Great Fire of 1666, Robert Hooke was appointed chief surveyor. His concern or purpose was with the rebuilding of the city. His interest in the elasticity and strength of substances complemented this concern and led him to carry out a variety of inquiries.

In general terms, the product of inquiry is conceptualized as a network of explanations that have an amazing power to explain and predict the way the natural world works. The precise nature of this network of explanations is difficult to conceptualize (and possibly contentious), but it's important to try because students do need to have some idea of what they are trying to learn. This curriculum presents three different types of explanation: generalizations that relate one pattern to another (e.g., wooden materials burn); generalizations that elaborate, often quantitatively, on a relationship (e.g.,

the wetter the wood, the slower it burns); and analogies and models that link generalizations and add an extra dimension of understanding (e.g., wood behaves as if it is a bundle of fibers stuck together).

d. The method of scientific inquiry and the assumptions on which it is based have evolved over the past 400 years.

Many portrayals of scientific inquiry are incomplete. This is especially true of those curriculum materials that emphasize the development of process skills. Many neglect the assumptions on which such inquiry is based.

For example, scientists assume that the world is regular and, in their experiments, give particular attention to reproducibility. But few curriculum materials, if any, give due attention to the idea of reproducibility (called *repeatability* in these materials). Without reproducibility, scientific inquiry as we know it would not exist. Reproducibility is a crucial test of a good experiment, yet it is ignored more often than not.

Many curriculum materials focusing on the process dwell heavily on the collection and manipulation of numerical data. But seldom do they ask, "Why do scientists use numbers?" or "Can good science be done without numbers?" These are crucial questions that also are ignored more often than not.

One further example: Seldom is it acknowledged that scientific research is undertaken by a community of scientists and that what counts as scientific knowledge is the consensus scientists hold about the best explanation of the way our world works—that knowledge, rather than being out there for the finding, is inside their heads. This widely accepted view of the nature of scientific knowledge is scarcely recognized in curriculum materials.

e. Scientific thinking is a refinement of everyday thinking.

This was Einstein's view, and these materials share this view. This is comforting, but it is nevertheless important to consider just how it is a refinement of everyday thinking.

The purpose of scientific activity is to produce a body of understanding that stands the test of time; it is about "knowing for sure." Scientists take pains to make sure they can trust their findings. This enables one scientist to build on and incorporate the findings of others in a common quest to understand how the universe works.

Scientific activity is careful and deliberate and sometimes probes areas of inquiry that everyday activity would ignore or take for granted. The carefully designed experiment is often a central element of this activity.

Scientific activity uses reductionist methods. It reduces complexity by systematically "pulling apart" a phenomenon and investigating each element in turn, with a view toward assembling a comprehensive understanding of the phenomenon.

Scientific activity is systematic and methodical. It is different from cooking, which uses trial and error.

Simple, elegant explanations and theories are preferred to convoluted ones, to which there are many exceptions and qualifications. For example, the early chemists put forward the idea of "phlogiston," an invisible flammable substance, to explain combustion. However, there were lots of problems with this theory: Sometimes phlogiston seemed to weigh a lot, and sometimes it appeared to weigh less than nothing. The puzzle was eventually solved by Lavoisier, who suggested that when something burned, it combined with the active part of air (oxygen). His simple and elegant theory explained everything and much more. Physicists today are on a quest to find a universal theory of gravitation and electromagnetic interactions.

Einstein's view of science as being a refinement of everyday thinking is also useful pedagogically. It suggests the goal of helping students refine or develop their everyday thinking capacity. To reach this goal, it would seem logical for the teacher to begin by revealing the students' everyday thinking about problem solving and then helping them construct an appreciation of how scientists solve problems. This pedagogical strategy—a constructivist strategy—is the theme of the next section.

Pedagogic Strategy

So how does one help students appreciate the art of solving scientific problems, become better scientific problem solvers, and gain an awareness of the way in which scientists have developed our understanding of natural phenomena? This was the central question that needed to be addressed when devising a suitable teaching strategy.

Central to the strategy finally adopted was the fairly obvious idea of actually challenging the students with meaningful problems to solve—problems they felt they could tackle with a view to the teacher subsequently helping them become better scientific problem solvers. These problems were selected after reviewing hundreds of students' projects in an attempt to identify their range of interests and the types of difficulty they encountered.

The strategy begins with the teacher introducing a problem. Considerable skill and sensitivity are needed to help the students find meaning in the problem, take ownership of it, and feel confident that solving the problem is within their capabilities. This step is important if the students are to accept the challenge of designing their own procedures. The students, in pairs or sometimes in small groups, are then given the task of finding a solution to the problem.

After they have attempted to solve a problem, the students are frequently provided with the opportunity to discuss how they tackled it. In doing so, they reflect on the procedures they used and make their procedures explicit. Making students aware of their thinking is an important precursor to their learning. Moreover, a teacher who elicits his or her students' ideas is in a powerful position to respond to their needs.

This then sets the stage for inviting students to compare how they set about solving the problem with how scientists might have tackled it. Finally, the students are encouraged to practice on a similar problem.

Typically, then, a lesson follows a four-stage sequence. Let us take an example. In Chapter 4, the lesson is "**Are Some Measures Better Than Others?" The lesson asks the question, "Which magnet is stronger?"

Stage 1: Attempting to Solve a Problem

The teacher introduces the problem by describing and maybe demonstrating the attempts of two students, Marie and Monique, to make magnets. After setting the scene and (hopefully!) gaining the students' interest, the teacher then challenges the students to help Marie and Monique find out which of two magnets is stronger. The students work in pairs. Each pair is provided with two magnets and access to assorted materials: paper clips, plotting compasses, thread, iron filings, rulers, elastic bands, and so on. As the students work, the teacher can gain some insight into their preliminary thinking.

Stage 2: Reflecting on Student Procedures

a. The students are asked to write to Marie and Monique with their recommendations about the best strength tests. Research indicates that having an audience for a report adds meaning to the task of writing up what they did.

b. They share and compare their solutions with others.

Here the students are implicitly invited to reconstruct their thinking when appropriate.

Stage 3: Comparing Student Procedures to Those Used by Scientists

A case study in which Marie and Monique visit a scientist is used as a vehicle for stimulating discussion about problems of measurement. Through this discussion, the students compare the procedures they devised with those favored by scientists. This leads to a consideration of the ideas of reliability and sensitivity.

Stage 4: Practicing

The students test the reliability and sensitivity of some of the procedures. There are many variations of this overall strategy. Here are three important ones:

a. In Stage 3, when possible, a case study of how a scientist tackled a similar problem is introduced. In this way, students learn about the experiments of people such as Galileo, Mendel, Lavoisier, Aristotle, Fleming, Redi, Newton, and Hooke.

b. In Stage 3, questions for discussion are often included so as to change reading from a passive experience to an interactive and thoughtful one. These questions are printed in italics and frequently found at the end of a paragraph. They provide an opportunity to take time out for a brief discussion.

c. Many teachers will use this book to help students undertake independent science projects and investigations, possibly with a view to participation in a science fair. Stage 4 frequently provides the opportunity for students to practice on the type of projects other students have actually chosen. Alternatively, if students are undertaking a project of their own (e.g., one of those described in Chapter 1, "Starting Points"), then Stage 4 provides an excellent opportunity for reviewing progress and applying the ideas encountered in the lesson.

This pedagogic strategy is based on the current constructivist view of learning, a learning theory that has gained ascendancy over the past 20 years to the point of being almost universally accepted (for example, it has been endorsed by Project 2061 of the AAAS). The strategy is similar to the one suggested for teaching scientific concepts in my recent book *Predict, Observe, Explain* (Haysom and Bowen 2010), which might prove to be a useful reference.

Inquiry and Student Independence

A Framework for K–12 Science Education (NRC 2012) has this to say about student independence:

> Students should have opportunities to plan and carry out several different kinds of investigations during their K–12 years. At all levels, they should engage in investigations that range from those structured by the teacher—in order to expose an issue or question that they would be unlikely to explore on their own (e.g., measuring specific properties of materials)—to those that emerge from the students' own questions. As they become more sophisticated, students also should have opportunities not only to identify questions to be researched but also to decide what data are to be gathered, what variables should be controlled, what tools or instruments are needed to gather and record data in an appropriate format, and eventually to consider how to incorporate measurement error in analyzing data. (p. 61)

In Canada, the *Common Framework of Science Learning Outcomes* (CMEC 1997) makes similar points.

Developing student independence involves the teacher in moving from a more structured classroom to a less structured one. It involves providing students with more and more opportunities to make choices. It involves the teacher in behaving less of an authority and more of a resource, who helps, facilitates and responds flexibly to the needs of students.

Some think that this can promote a more enjoyable classroom, lead to greater student involvement and motivation, and improve a student's self-confidence. Others believe that if teachers relinquish some of their academic authority, this might create classroom-management problems. Paradoxically, some students might resist being offered more freedom, preferring instead the security that structure provides. Either way, it is easier said than done. Nevertheless, it might be possible for a teacher to progressively offer the students more freedom of choice as the students gain confidence and determination in solving their own problems. It could be worth trying, and Table 1 might be of some use to those who wish to check the amount of structure provided in the various stages of inquiry.

Introduction

Finally, the materials themselves are designed to facilitate the change from more structure to less. For example, the materials often invite discussion of how scientists have tackled a problem, a strategy that provides the teacher with the opportunity to distance herself from authority. Here is another example: In the Back to the Project items that conclude most of the lessons, there is often no need for the teacher to provide closure. Instead, the ideas that have been introduced can simply be allowed to continue to challenge the students. As was mentioned previously, the materials use a constructivist pedagogy, and one of the fundamental tenets of constructivism is that students are responsible for their own learning. It might be possible for a teacher to gradually offer the students more freedom of choice if the materials are used over three consecutive years.

Table 1: Stages of Inquiry and Increasing Student Independence

Stage of Inquiry	Teacher-Structured	Semi-Structured Increasing Student Independence	Independent Student
Idea for Inquiry (see Chapters 1 and 12)	The teacher assigns the Starting Point.	Student chooses from a list of Starting Points provided by the teacher.	Student identifies his or her own Starting Point.
Clarification of Idea or Preliminary Exploration (see Chapter 3)	The Starting Point is clarified by the teacher, who formulates questions for investigation.	The teacher helps clarify the problem by discussing it and asking questions.	The student carries out preliminary exploration, formulates questions, and makes decisions about a course of action to follow.
Design of Investigation (see Chapters 4, 5, and 6)	The teacher provides the student with a plan to follow, measurement procedures to use (if necessary), and variables to control.	The teacher indirectly assists the student with the design of the investigation and discusses problems of measurement and the control of variables.	The student designs the procedure for the investigation.
Collection of Data (see Chapter 7)	The teacher directs the student to check for reproduceability, to search for errors, and so on.	The teacher asks about reproduceability, sources of error, and so on.	The student independently checks for reproduceability and searches for errors.
Analysis and Interpretation of Data (see Chapters 8 and 9)	The teacher tells the student how to analyze the data and interprets the results.	With guidance from the teacher, the student tries to make sense of his or her results.	The student analyzes data and considers their meaning and significance.
Presentation of Investigation (see Chapter 10)	The teacher outlines how the student should present inquiry.	The teacher discusses alternative ways to present the inquiry.	The student decides how to present the inquiry.

Introduction

Structure of the Curriculum

The curriculum begins with an invitation to the students to choose a practice project, a Starting Point (Chapter 1), and culminates with the students choosing a science fair project of their own (Chapter 12, "Generating Ideas for Projects").

Chapter 2, "An Overview of the Nature of Scientific Inquiry," provides an orientation to the way in which scientists approach a problem and a perspective on the nature of scientific inquiry, which is the practice of scientists.

The remaining 10 chapters each focus on a different aspect of scientific investigations—different scientific practices. I anticipate that as students carry out their warm-up projects (or projects of their own choosing), they will encounter problems similar to those introduced in these chapters. The chapters provide indirect help along the way. They are sequenced along a timeline from having an idea for an inquiry (Chapters 1 and 2) to clarifying the idea (Chapter 3) to designing an investigation (Chapters 4, 5, and 6) to collecting data (Chapter 7) to analyzing and interpreting data (Chapters 8 and 9) and, ultimately, to presenting the findings. This sequence, although seemingly logical at first glance, is frequently not followed in practice. Indeed, scientists seldom follow this process. It certainly isn't carved in stone, and teachers may well decide to change the order to help students with the problems they encounter.

Each of the three student books in the series follows this sequence of chapters, and the books progressively increase in complexity and difficulty as one proceeds from the first book to the last one. The ideas in the later books build on those presented in the earlier books and make greater cognitive demands on the students. Let us take a couple of examples.

Example 1

Chapter 5, "Variables and Their Control": This chapter, building on students' natural idea of fairness, introduces the idea of a fair test through the control of variables in the grades 5–8 book. Once a fair test has been achieved, then it is possible to move on to the grades 7–10 book to find out how one variable depends on another. Finally, in the grades 8–12 book, students are invited to examine the rather demanding problem of designing a boat hull, a project in which there are many interacting variables.

Example 2

Chapter 7, "Sources of Error": This chapter deals with sources of error. The grades 5–8 book focuses on errors of measurement and taking the average. The grades 7–10 book examines the problems of tracking down sources of error that arise from uncontrolled variables, and the grades 8–12 book deals with variables that one cannot control, such as those encountered when working with human subjects, animals, and plants. In the grades 8–12 book, Chapter 7 introduces statistics and sampling.

In the series, the lessons in the first book (grades 5–8) are marked with a single star (*), the lessons in the next book (grades 7–10) with two stars (**), and those in the final book (grades 8–12) with three stars (***). The series has been field-tested with middle school students—the one-star lessons with grade 6, the two-star lessons with grade 7, and the three-star lessons with grade 8.

If a teacher has access to all three books, the teacher would have extra flexibility to mix lessons to match the needs of his students. Indeed, a comprehensive course for mature students (grades 9–12) on the practices of scientists might use this lesson sequence: 1***, 2*, 2***, 3 (all lessons), 4 (all lessons), 5 (all lessons), 6 (all lessons), 7 (all lessons), 8**, 8***, 9 (all lessons), 10**.

Introduction

Content Overview

Chapter	Progression, Level of Sophistication		
	Grades 5–8 Book (*)	Grades 7–10 Book (**)	Grades 8–12 Book (***)
1. Starting Points	• Paper Helicopters • What Makes Seeds Grow? • Etc. (2)	• Louis Braille's Invention • Archimedes' Pump • Etc. (2)	• Smoking Chimneys • Acid Rain and Pollution • Etc. (2)
2. An Overview of the Nature of Scientific Inquiry	• Beginner Scientists and Experienced Scientists (2)	• What Do Scientists Do? (3)	• Simplifying Complex Scientific and Technological Problems (2 or 3)
3. Science Without Numbers	• Wondering Why (2)	• Looking for Similarities and Differences (3 or 4)	• Searching for Patterns (4)
4. The Numbers Game	• Learning to Play the Numbers Game (2)	• Are Some Measures Better Than Others? (3)	• Designing Your Own Measures (4)
5. Variables and Their Control	• Being Fair (1)	• Finding Out How Much It Matters (2)	• Isolating Variables: Reducing Complexity (2)
6. Experiment Design	• Getting Experiments to Work: Repeatability (2)	• Designing Good Apparatus (2)	• Preparing Experiment Designs (1 or 2)
7. Sources of Error	• Taking the Average (1 or 2)	• Searching for Errors (1 or 2)	• Sampling: An Introduction (2)
8. Making Sense of Your Results	• Charting Your Data (2)	• Graphing Your Data (1 or 2)	• Interpreting Graphs (2)
9. Explanations	• Looking for Patterns and Trends: Generalizing (2 or 3)	• Getting Explanations That Fit (5)	• Deepening Your Understanding: Analogies and Models (3)
10. Sharing Your Findings	• Displaying Your Project (2)	• Writing Your Report (2)	• Talking About Your Project (1)
11. Judging Projects	• Checking for Quality (1)	• Suggesting Improvements (1)	• Making Judgments (1)
12. Generating Ideas for Projects	• Ideas From Previous Science Fairs (1)	• Ideas All Around You (1)	• Ideas From the Scientific Literature (1 or 2)

Note: The numbers in parentheses give an estimate of the number of 40-minute periods required to complete the lesson.

Be Safe!

It is important to set a good example and remind students of pertinent safety practices when they perform an experiment.

1. Always review Material Safety Data Sheets (MSDS) with students relative to safety precautions.

2. Remind students to only view or observe animals and not to touch them unless instructed to do so by the teacher.

3. Use caution when working with objects such as scissors, razor blades, electrical wire ends, knives, or glass slides. These items can be sharp and may cut or puncture skin.

4. Wear protective gloves and aprons (vinyl) when handling animals or working with hazardous chemicals.

5. Wear indirectly vented chemical-splash goggles when working with liquids such as hazardous chemicals. When working with solids such as soil, metersticks, glassware, and so on, safety glasses or goggles should be worn.

6. Always wear closed-toe shoes or sneakers in lieu of sandals or flip-flops.

7. Do not eat or drink anything when working in the classroom or laboratory.

8. Wash hands with soap and water after doing the activities dealing with hazardous chemicals, soil, biologicals (animals, plants, and so on), or other materials.

9. Use caution when working with clay. Dry or powdered clay contains a hazardous substance called silica. Only work with clean wet clay.

10. When twirling objects around the body on a cord or string, make sure fragile materials and other occupants are out of the object's path.

11. Use only non-mercury-type thermometers or electronic temperature sensors.

12. When heating or burning materials or creating flammable vapors, make sure the ventilation system can accommodate the hazard. Otherwise, use a fume hood.

13. Select only pesticide-free soil, which is available commercially for plant labs and activities.

14. Many seeds have been exposed to pesticides and fungicides. Wear gloves and wash hands with soap and water after an activity involving seeds.

15. Never use spirit or alcohol burners or propane torches as heat sources. They are too dangerous.

16. Use caution when working with insects. Some students are allergic to certain insects. Some insects carry harmful bacteria, viruses, and so on. Wear personal protective equipment, including gloves.

17. Immediately wipe up any liquid spills on the floor—they are slip-and-fall hazards.

Lesson Plans

The remainder of this guide provides the details of individual lessons: purpose, time allocation, apparatus, materials, and suggested approach.

The suggested approach is, of course, explanatory rather than prescriptive; it explains why the student resource books were written in this particular way. Each component activity in the suggested approach is related to the student books. The apparatus and materials required have been kept to a minimum. Nothing fancy is needed.

Finally, a word about purpose: An attempt has been made to make this more meaningful than tersely stated educational objectives. Moreover, the descriptions of individual lessons in each chapter are preceded by a short overview of ideas about the nature of science and scientific inquiry, on which the lessons are based. To put these ideas about the purpose of the lessons into a broader perspective, extracts from the *National Science Education Standards* (NRC 1996) and *A K–12 Framework for Science Education* (NRC 2012) have been provided. The extracts from the Framework often provide a valuable elaboration.

Chapter 1:
Starting Points

All students, regardless of whether or not they have previously experienced a science fair, are invited to choose a Starting Point. The idea is for them to be given time to investigate these Starting Points over the course of the next 20–30 lessons.

As students work on a project, they will undoubtedly meet all sorts of problems: formulating "good" (scientific) questions, measuring, designing good apparatus, devising good experiments, graphing, making sense of what they have learned, and so on. But they can get help along the way from the experiences in these books, from the teacher, and from their friends and classmates. The problems they encounter can be shared in an open-forum discussion.

The books look at scientific ways of dealing with these problems, and toward the end of each lesson it is important for the students to take time out to think about how they might improve their projects and apply the ideas they have learned to their own situation.

At the same time, those students who have not experienced a science fair will get an idea about what the fairs are like, and at the end of the course the students have the opportunity to choose a project of their own. Those who have had previous experience might prefer to work on a project of their own choice instead of one of the Starting Points.

There are 15 Starting Points altogether, but of course the list of possibilities is endless and

teachers might prefer to devise their own. Those marked with one or two asterisks are simpler and more suitable for introductory or intermediate courses. Some of the Starting Points address physics problems, some chemistry, and others biology. Some have an engineering flavor, as scientists and engineers share similar practices.

Here is an extract from the *National Science Education Standards* (NRC 1996):

> Identify questions that can be answered through scientific investigations. Students should develop the ability to refine and refocus broad and ill-defined questions. An important aspect of this ability consists of students' ability to clarify questions and inquiries and direct them toward objects and phenomena that can be described, explained, or predicted by scientific investigations. (p. 145)

Here is an extract from *A Framework for K–12 Science Education* (NRC 2012):

> Engineering and science are similar in that both involve creative processes, and neither uses just one method. (p. 46) … [T]he goal of science is to develop a set of coherent and mutually consistent theoretical descriptions of the world that can provide explanations over a wide range of phenomena. For engineering, however, success is measured by the extent to which a human need or want has been addressed. (p. 48)

Chapter 1

1* STARTING POINTS

Purpose

The purpose of this lesson is twofold:

1. Give students a quick orientation to science projects and to the way the book is structured.

2. Get students started on a project.

Time Allocation

Two 40-minute periods

Apparatus and Materials

Slides or videos of previous science projects or fairs might provide a valuable orientation for students who have not had any previous experience.

Select about five Starting Points that you think will be of particular interest to your class. Assemble three Starter Packs for each Starting Point. The apparatus and materials required for each Starter Pack are listed below. The total number of Starter Packs (store them in trays) is 15—enough for 30 students working in pairs. This is sufficient to offer most normal-size classes some choice.

As they progress with their projects, the students will probably ask for other materials. It would be helpful to have a supply of "junk" materials and a simple tool kit available. Students should be encouraged to seek out their own materials whenever possible.

Before beginning, you should review and model the "Be Safe!" section at the beginning of the students' books (see also the "Safety in the Classroom" section in this book). You might consider asking students and parents or guardians to read and sign a safety acknowledgment form.

Suggested Approach

1. Read and discuss with students the first two sections of Chapter 1 of the grades 5–8 book ("Science Projects and Science Fairs: What Are They About?" and "Getting Started"). The introductory paragraph and cartoon are designed to encourage students to contribute what they already know about science fairs. This might be followed by showing them slides or videos of previous science fairs.

2. Introduce the students to the structure of the book. The book begins with a series of suggestions for getting started. The suggestions are followed by a series of lessons designed to help students solve problems they encounter along the way. This section concludes by helping students choose a project on their own. Refer the students to the contents pages and discuss the match between these and the structure of the book.

3. Briefly demonstrate each of the Starting Points. Briefly discuss the scope of each Starting Point with the students.

4. Organize the students in pairs or groups of three. Invite them to choose a Starting Point.

5. Invite the students to explore the scope of their chosen project by "playing around with it." Refer to the section "Let's Go!" They should compile a list of questions that interest them regarding their projects.

6. Discuss with the students (as a class or in groups) their questions. Invite them to design and carry out a pilot experiment to find out the answer to one of their questions. They

should be prepared to share their findings with others.

The students should be encouraged to continue working on a project they have chosen until it is complete. This is especially important if it is the first project they have undertaken. There are many ways to organize this; here are two examples:

- Students work at home, and the teacher sets aside class time for progress reports.

- The teacher sets aside class time once per week or every two weeks for students to continue their work.

Notes

Some of the trial teachers decided to do a science fair project alongside their students. One teacher who had access to a vacuum sandwich sealer decided to investigate Starting Point 14, "Life on the Moon." She "planted" mustard and radish seeds in Mason jars on a substrate of clay balls soaked in hydroponic solution. Each jar was placed in a plastic bag, and the bag was evacuated.

From time to time, she reported on her progress and initiated discussions with the students on what to do next. Although she admitted to occasionally trying to influence her students, she let them have the last word. This didn't inhibit her from "arguing like crazy," though!

She judged the experience as very worthwhile. There were, however, all sorts of problems:

- Seeds didn't germinate, although growing plants continued to thrive.

- There were arguments about control experiments.

- The students noticed that the plants generated their own atmospheres (the bags swelled).

- One experiment ran out of hydroponic solution.

- Students suspected that a bag leaked in another experiment.

As a culmination of the project, the students challenged her to enter the science fair.

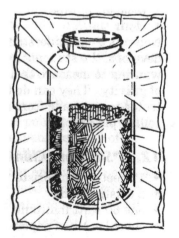

Chapter 1

STARTER PACKS

1. *Paper Helicopters (and Paper Airplanes)

- Ruler
- 3 sheets of paper
- Masking tape
- Scissors
- 3 note cards
- 50 paper clips
- 5 straws

(A blackline master for the helicopter illustrated in the diagram is provided in Appendix A.)

2. *What Makes Seeds Grow?

- Packet of radish seeds
- Packet of bean or pea seeds
- 6 glass jars or clear plastic cups
- Paper towel

3. *Check Rust

- 6 test tubes
- Steel wool
- 6 nails
- Can of oil
- Small salt packet

4. *What Makes Sow Bugs Move?

- Wood lice/sow bugs: You can ensure a good supply of wood lice by establishing a "farm" in the fall. This simply requires a large ice cream container, 5 cm of soil, and some rotting leaves and wood. Keep the soil damp.
- Small bottles of apple juice, orange juice, and vinegar
- Filter paper and medicine droppers (useful, but not required)

5. *Bouncing Balls

- Bouncy balls of different sizes
- Glass balls (marbles) of different sizes
- Meterstick
- Piece of rubber and piece of wood, each about 20 cm^2

6. **Louis Braille's Invention

- Sample of Braille (from the National Institute for the Blind)
- 10 index cards
- Ballpoint pen
- Piece of soft wood

7. **Archimedes' Screw

- Bowl
- Beaker or plastic ice cream container
- Retort stand/clamp/boss head, or 6 bricks
- 2 broom handles of different thickness
- Two 2 m lengths of tubing (different diameters)

- Masking tape
- Measuring cylinder (100 ml)

8. **Electric Cells

- Lemon juice
- Apple juice
- Copper, zinc, iron foil, or nails
- Beaker or glass jar
- Connecting wire and clips
- Insulated copper wire
- Compass
- Flashlight bulb

9. **The Right Nail for the Right Job

- 2 short (about 25 cm) pieces of wood
- Small pieces of fiber board and polystyrene foam (if possible)
- 3–6 different sizes of common nails (about 3 of each size)
- Hammer
- Strong nylon fishing line
- Ruler
- Bucket and sand or weights

10. **Suffocating Candles

- 2 candles
- Book of matches
- 6 jars or beakers of different shapes and sizes (200–1000 ml)

- 8 plywood blocks (approx. 5 cm × 5 cm × 1 cm)
- Measuring cylinder

11. ***Women Can! Men Can't!

- Measuring tape
- Chair

Note: As their understanding of the problem develops, the students may find themselves wanting to measure their "balance point" (centers of gravity). They can do this using an improvised seesaw made from a plank of wood and a rolling pin or log.

12. ***Smoking Chimneys

- 1 sheet of thin bristol board
- 2 elastic bands
- 60 W lightbulb in holder
- Small piece of paper

13. ***Acid Rain and Pollution

- Sample (1 L) of freshwater from local source (*Note:* The sample should contain a variety of microscopic organisms. Collect some mud and plants as well.)
- Microscope and slides
- Medicine dropper
- 6 test tubes
- Measuring cylinder
- M/1,000 nitric acid and/or sulfuric acid (100 ml)

14. ***Life on the Moon

This could be a whole-class project. Its scope is broad. The apparatus and materials required may seem rather extensive, but they are much less so when seen in the context of at least one month's experimental work for a whole class.

- Vacuum pump (water powered) capable of reducing pressure to the point at which water boils. This is the most expensive piece of equipment, but the cost is not exorbitant.

- Supply of seeds and other organisms: Pea seeds, baker's yeast, mustard seeds, and mung bean seeds (available from health food stores) will provide enough variety to begin.

- Plant nutrients: sodium bicarbonate, sugar, hydroponic nutrient (available from garden suppliers), and, if possible, inert plant aggregate.

- Miscellaneous: a generous supply of hard-glass boiling tubes and one-holed rubber stoppers to fit, glass tubing and rod, vacuum grease and sealing wax, pressure tubing, and simple screw clamps

- Additionally, you should expect the students to request other apparatus, such as measuring cylinders, storage bottles and medicine droppers. It's not possible to anticipate all their needs, which might include such items as fluorescent lamps.

- Not all the experiments students suggest need to be carried out in a vacuum. For example, the effects of varying light periodicity and gravitational force do not require a vacuum, at least not in the first place. Similarly, students do not need a vacuum to find out whether plants can derive their carbon dioxide from sources such as sodium bicarbonate, but they do require a carbon-dioxide-free atmosphere (soda lime is a good absorbent).

15. ***Falling Leaves

- Ruler
- Scissors
- 3 index cards

Note: The students may ask for a stopwatch as their inquiry develops.

Chapter 2:
An Overview of the Nature of Scientific Inquiry

There are many possible answers to the question, "What is the nature of science and scientific inquiry?" Words alone are not enough to provide the answers. Yet it's important for students to begin to see their projects from the perspective of real scientific activity. This chapter provides students with experiences through which they can understand what science is all about and the sort of demands a scientific investigation makes on them.

There are, of course, many types of scientific inquiry. Science, technology, and engineering differ from one another. Simply put, science—for example, the study of the behavior of animals in their natural environment—aims to add to our understanding of the natural world. On the other hand, technology—for example, the formulation of better house paints—aims to apply our understanding gained from science (in this case, our chemical understanding of polymeric reactions) to a practical problem. Engineering—for example, the design of a turbine to harness tidal power—seeks to address important human needs. The boundary between technology and engineering is a fuzzy one, and some would claim they are the same. Such studies nevertheless do have significant features that distinguish them from the more casual, undisciplined inquiry of the layperson. They all use similar types of inquiry. Many examples of each type of inquiry are featured in the students' texts.

The first lesson invites the students to contrast everyday common sense with science sense. The students will probably notice how careful and precise scientists are and begin to see why science is often carried out in laboratories with specialized equipment.

The second lesson emphasizes the empirical—the place of contrived experiments and systematic observation and how scientists use them to unravel problems.

The third lesson introduces engineering as a science concerned with solving practical problems. At the same time, it attempts to show how scientists and engineers use the strategy of reductionism to break down complex problems. Whereas the process of trial and error provides information about specific cases, reductionism leads to an understanding about the components of the general case.

The three lessons draw attention to the methods of investigation scientists use and the purpose of scientific activity.

Here is an extract from *A Framework for K–12 Science Education* (NRC 2012):

> Thus a common elementary school activity is to challenge children to use tools and materials provided in class to solve a specific chal-

lenge, such as constructing a bridge from paper and tape and testing it until failure occurs. Children's capabilities to design structures can then be enhanced by having them pay attention to points of failure and asking them to create and test redesigns of the bridge so that it is stronger. (p. 70)

2* BEGINNER SCIENTISTS AND EXPERIENCED SCIENTISTS

Purpose

This lesson is designed to provide the students with a broad view of the nature of science and scientific inquiry. It provides them with an overall perspective in which to see the work they are doing on their projects. The first part of the lesson is designed to help them distinguish between everyday sense-making and disciplined sense-making.

Time Allocation

Two 40-minute periods

Apparatus and Materials

None required

Suggested Approach

1. Ask the students what they know about puddles drying up. Read the introduction and discuss it.

2. Introduce the Exploration and ask the students to devise an experiment to solve one aspect of the puddle problem. After about 10 minutes, review one or two answers.

3. Ask the students to dramatize the conversation between the two experienced scientists, Michelle and Joanne.

4. Identify and discuss the differences between the beginner and experienced scientists.

5. Have students form small groups and challenge them to refine their own experiments. After about 10 minutes, review answers from one or two groups.

6. Provide the opportunity for students to review and refine some of the experiments they have designed in their projects (possibly for homework).

7. The remainder of the lesson attempts to help the students grasp how scientists build scientific understanding. Read the section What Have Scientists Discovered About Puddles? and pause for discussion when appropriate.

2** WHAT DO SCIENTISTS DO?

Purpose

This lesson is designed to help the students increase their understanding of the nature of science and scientific inquiry. They compare their attempts to solve an apparently simple problem with those of scientists.

Time Allocation

Three 40-minute periods

Apparatus and Materials

- Golf balls and table tennis balls
- Pieces of foam or carpet
- 2 packets of cards
- Several pieces of paper
- Stapler (may be necessary)

Suggested Approach

1. After a brief discussion of the nature of science and scientific inquiry, ask the students to write down their definitions of *science* and *scientific inquiry*.

2. Introduce the Exploration and set the students to work (they might try to make sense of their findings for homework).

3. Read about Aristotle's and Galileo's ideas. Pause to discuss who the students believe was right. Continue reading, pausing to discuss the italicized questions. You can find useful videos of the penny and feather experiment and the experiment using a strobe on YouTube (*www.youtube.com/watch?feature=endscreen&v=clom4DdnFfM&NR=1*; *www.youtube.com/watch?v=xQ4znShlK5A*).

4. Review the way science has helped us understand the way objects fall. After discussing the ideas students have about science, invite them to refine their original definitions.

Chapter 2

2*** SIMPLIFYING COMPLEX PROBLEMS

Purpose

The purpose of this lesson is to help the students appreciate how complex problems such as Smoking Chimneys, Acid Rain and Pollution, and Life on the Moon can be simplified to make them susceptible to systematic inquiry. Many technological and engineering problems can be addressed in this manner. The rationale offered is based on a critique of the process of trial and error.

Time Allocation

Two or three 40-minute periods

Apparatus and Materials

- Scales

- Sand (about 1 L)

- 2 ice cream containers (1 L each)

- String

In addition, each group of students will require a tube of quick-setting model airplane glue or 1 m of masking tape, 6 sheets of paper, and a pair of scissors.

Suggested Approach

1. Introduce the bridge-building competition.

2. Divide the class into teams and allow them about 30 minutes to build their bridges.

3. The students are asked to predict which bridge will win and why.

4. Test the bridges.

5. Discuss the competition, asking students how they might improve their bridges. They will intuitively begin to look for features of or patterns among the designs and speculate about how these are related to the bridge's strength. Ask them what they would do if they had a week to prepare for a second competition. This discussion provides the students with the opportunity to make explicit their ideas about the processes of scientific inquiry.

6. Read the section "Two Ways to Improve Bridge Design," and invite the students to comment on the ideas.

7. Invite each student team to investigate one aspect of bridge design ("Be an Engineer") and share his or her findings with the class. If time allows, the class might like to use the understanding gained to build an even better bridge.

8. Invite the students to reflect on the possibility of simplifying their projects.

Chapter 3:
Science Without Numbers

Scientific inquiry often proceeds qualitatively, especially at the early stages of an exploration into a new phenomenon: The scientists' goal seems to be to map what they perceive to be the significant features of the phenomenon, with a view to sorting through and understanding the phenomenon. This chapter highlights three exploration strategies that scientists use.

1. To provide direction to what otherwise might be random exploration, researchers often formulate questions that focus their attention.

2. Scientists categorize (e.g., conductors or nonconductors, herbivores or carnivores). They search for similarities and differences in the behavior of things, perceiving what they consider important, relevant, and interesting patterns of the behavior they are studying. Their perceptions act as organizers for their observations.

3. Scientists then try to organize these perceptions or patterns or categories, associating one with another (e.g., wet objects don't burn, metals conduct electricity). Sometimes this is called hypothesizing. This is followed by searching for further explanation, either in the form of a network of associated patterns or in trying to understand why two patterns are related. Scientists assume that the world is like a machine: It is regular, and their task is to discover just how it works.

These strategies of scientists do not necessarily require the use of numbers.

Here is an extract from the *National Science Education Standards* (NRC 1996):

> Identify questions that can be answered through scientific investigations. Students should develop the ability to refine and refocus broad and ill-defined questions. An important aspect of this ability consists of the students' abilitiy to clarify questions and inquiries and direct them toward objects and phenomena that can be described, explained, or predicted by scientific investigations. (p. 145)

Here are extracts from *A Framework for K–12 Science Education* (NRC 2012):

> Scientists and engineers investigate and observe the world with essentially two goals: (1) to systematically describe the world and (2) to develop and test theories and explanations of how the world works. In the first, careful observation and description often lead to identification of features that need to be explained or questions that need to be explored. (p. 59)

> By grade 12, students should be able to analyze data systematically, either to look for salient patterns or to test whether the data are consistent with the initial hypothesis. (p. 62)

Chapter 3

3* WONDERING WHY

Purpose

Students often experience difficulty when starting a project. Their exploration of a phenomenon tends to remain narrow in scope, and their "messing about" seems to lack direction or purpose. This lesson is designed to help add purpose to students' explorations by encouraging them to make their "casual wondering" explicit and talk about their thinking.

Time Allocation

Two 40-minute periods

Apparatus and Materials

Each pair of students will require the following materials:

- A template from which they can cut out the standard paper helicopter (See blackline master in Appendix A.)

- Scissors

- Ruler

- Paper and cardstock

- Paper clips

Suggested Approach

1. Read or talk about the first two paragraphs of Starting Point 1 and invite the students to find out all they can about this topic. This part of the process can occupy one period, especially if the teacher prompts students to make their thinking explicit: "What are you doing now?" "What do you think might happen?" "Does/did that make sense to you?"

2. Review some of the charts in which students have recorded their thoughts and actions.

3. Read the section "Scientists Ask Questions" and generate from the class a list of questions they have asked about paper helicopters.

4. Read about the project on crickets. Individuals or pairs of students then formulate questions worth asking. They then select a question and design an experiment to answer it.

5. For homework, invite students to explore one of the projects from "Back to the Project" or a project of their own.

Alternative or Optional Activity

A charming alternative to the exploration with paper helicopters is to give students the task of growing the tallest bean or pea in the class. In the week before the lesson, read and discuss Starting Point 4. Give each student two or three seeds to take home for experimenting. For each seed, ask students to record information in the following categories: What I Did, Why I Did It, and What I Saw.

The students will likely be interested in seeing one another's plants, so you can allow them 5–10 minutes to do so.

Begin the discussion by asking students what they saw (the seed swelling, the shell cracking, the root appearing, and so on). Write their answers on the board. These are the observations they connect with growth.

Scientists would likely proceed in this inquiry by trying to associate (tentatively) growth with patterns of treatment (conditions). You might encourage students to discuss this together under three headings:

- Conditions that seem to speed up growth

- Conditions that seem to slow growth

- Conditions that don't seem to matter or with effects about which we disagree

This is the beginning of science sense-making. The next step is to design experiments to test each of the ideas.

Note: We do things. We see things happen. We try to make sense. The sense we make directs our future actions and structures our observations. In the everyday world, this happens spontaneously. In this lesson, however, we are trying to help students begin reflecting on the process—an enormously difficult but important task.

3** LOOKING FOR SIMILARITIES AND DIFFERENCES

Purpose

There are many instances in the history of scientific discovery that suggest that in the early stages of inquiry, scientists often proceeded by

- observing interesting similarities and differences in the behavior of things,

- trying to relate these patterns of behavior in their search for order within them, and

- doing experiments to test their tentative explanations.

The scientific method is often portrayed as one in which observation is followed by hypothesis and hypothesis by experiment. However, it is recognized that, in practice, the interplay between observation, hypothesis, and experiment may well be more complex and dynamic than this method.

This pattern of thinking is particularly evident following the birth of interest in a new topic,

and awareness of the pattern may help students at the beginning of their projects. This lesson is designed to enable the students to experience and reflect on the process.

Time Allocation

Three or four 40-minute periods

Apparatus and Materials

Each group (pair) of students will need the following:

- Candle

- Tongs or clothes peg

- Aluminum foil (about 30 cm square)

- Water in a medicine dropper

- White vinegar in a medicine dropper

Chapter 3

- A few (e.g., 3) cream pots or test tubes
- A few craft sticks
- Hand lens (optional)
- Containers holding four white powders: F (plaster), R (starch), L (baking soda), and E (mixture of starch and baking soda)

Suggested Approach

1. Introduce the spaceship simulation and set up the first task.

2. During the next period, review the students' findings and consider the first two questions in "Questions for Discussion." When they chart their results, the students might finish with a table such as the one below.

3. Set up the second task using Sample E.

4. In the next class period, review students' findings and consider the third of the "Questions for Discussion."

5. The section "The Beginnings of Chemistry" is designed for interactive reading. The italicized questions inserted in the text are designed to stimulate full-class discussion. You might begin by demonstrating the reaction between malachite (use copper carbonate or copper hydroxide) and carbon. You might demonstrate the experiment later with the burning candle.

6. In the section "Back to the Project," the shape of a pile of sugar may well be worth demonstrating. It's simple to do and illustrates the various aspects of scientific thinking highlighted in this lesson. The shape (sometimes called the angle of repose) depends on the material, the height of drop, and other factors.

Sample	Water	Vinegar	Heat
F (plaster)	Dissolves, then goes solid	Dissolves, then hardens	Turns gray, smokes a bit
R (starch)	Thick like glue, then hardens	Thick like glue, then hardens	Smokes, turns brown
L (baking soda)	Dissolves	Fizzes and disappears	Smokes a bit, stays the same

3*** SEARCHING FOR PATTERNS

Purpose

When scientists try to make sense of the bewildering variety of natural phenomena, they attempt to discover the natural order of things. They often begin by searching for patterns and progressively seek to tease out the underlying patterns they assume exist. This lesson focuses on the process of searching for patterns.

Time Allocation

Two or three 40-minute periods

Apparatus and Materials

None required

Suggested Approach

Part 1

1. Introduce and read "Dave's Science Fair Project."

2. Have the students form small groups, and ask them to make whatever sense they can of Dave's data. Record the patterns they observe on the board.

3. Read aloud the fable "A Lost Child Keeping Warm." Ask students how they think the fable might continue. Then invite them to suggest how Dave might develop his project.

Part 2 (optional, for students who want to be challenged)

4. Introduce "Plant Pattern Puzzle." This is a completely original problem. Read the first two sections, pausing to discuss the italicized questions.

5. This should prepare the class for a small-group assignment in which they attempt to solve the "Plant Pattern Puzzle." It's a difficult assignment, but you can help by encouraging students to list the patterns they perceive in the data.

6. Review the students' answers, first listing the patterns they perceived. If any students have been successful in solving the puzzle, invite them to tell how they think the patterns are related. If not, introduce the idea that the color of the plant is determined by three factors—a red factor, a blue factor, and a yellow factor (see solution below).

7. Compare the scientific process with the process of decoding secret messages.

8. Read about and discuss Mendel's work. Invite the students to compare the "Plant Pattern Puzzle" with Mendel's experiments and findings.

Part 3

9. If any students have been working on the "Falling Leaves" project, ask them to report on their progress. If no students are working on this project, repeat the demonstration and invite discussion.

10. Invite the students to discuss in small groups how they would tackle one of the projects outlined at the end of the section.

Solution to Plant Pattern Puzzle

Here are some patterns:

a. All red plants have the same number of petals and the same number of leaves (the same holds for other colors).

b. The number of leaves and the number of petals are the same.

c. The leaves are arranged in two clusters. Each cluster contains one, two, or three leaves.

d. Only the red, blue, and yellow plants give the original color when bred with a flower of the same color.

e. When red, blue, and yellow plants are interbred, just one color is obtained (that of the mix).

f. Add another pattern observed.

The Way the Patterns Are Interrelated (the Code)

a. The color (and leaf clusters) of a plant is determined by two of the following factors: a red (one leaf) factor, a blue (two leaf) factor, a yellow (three leaf) factor.

b. The color of the plant and its leaf clusters result from combining two factors. Color factors combine or blend in the same way as paints do when mixed: Red and blue make purple, yellow and blue make green, and red and yellow make orange.

c. When plants are bred, there is an exchange of factors. Each plant contributes one color and leaf factor to the seed (offspring).

Chapter 4:
The Numbers Game

Have you ever asked yourself why scientists use numbers? It's an intriguing question. Science and mathematics appear to walk hand in hand. Why is this the case? Rather than there being a single reason, it seems that there are many reasons and that they are interrelated.

1. In the first place, numbers enhance our powers of description. They add precision to our observations and extend our range. They also add intersubjectivity—some would say objectivity—to our observations, thus facilitating communication within the scientific community by enabling scientists to share their data. This chapter, which introduces the numbers game, highlights these reasons. The chapter begins by helping the students appreciate the need to measure and then proceeds to introduce, in simple ways, a variety of important ideas (e.g., validity, reliability, sensitivity, units of measurement, standards, calibration).

2. Science seeks to describe the relationship between different aspects of a phenomenon. Newton and Leibnitz invented calculus to do just this. They brought in the ideas of tangent, acceleration, slope, differential, and the infinitesimal. Chapter 9 (on the interpretation of data) portrays mathematics as a language—the language of science. Indeed, Pythagoras had called it the language of nature.

3. Since Newton's time, mathematics and

science have become even more intimately interrelated. Mathematics has become a way of thinking about nature, and even of conceptualizing its workings. Scientists assume that nature is regular and mathematics has the power to describe, interpret, and communicate an understanding of that regularity. This idea is introduced in Chapter 9, "Explanations."

Here is an extract from the *National Science Education Standards* (NRC 1996):

Design and conduct a scientific investigation. Students should develop general abilities, such as systematic observation, making accurate measurements, and identifying and controlling variables. (p. 145)

Here is an extract from *A Framework for K–12 Science Education* (NRC 2012):

Decisions must also be made about what measurements should be taken, the level of accuracy required, and the kinds of instrumentation best suited to making such measurements. As in other forms of inquiry, the key issue is one of precision—the goal is to measure the variable as accurately as possible and reduce sources of error. The investigator must therefore decide what constitutes a sufficient level of precision and what techniques can be used to reduce both random and systematic error. (pp. 59–60)

Chapter 4

4* LEARNING TO PLAY THE NUMBERS GAME

Purpose

The purpose of this activity is to help students appreciate the need to quantify (use numbers).

- Our senses are not always sharp enough.

- Our senses can deceive us.

- Numbers can help us compare more than two things.

- It's important to clarify what exactly we are measuring (the validity of the measurement).

- Numbers can help us identify how much we have of something.

Time Allocation

Two 40-minute periods, plus homework requirement

Apparatus and Materials

You will need about three sets of apparatus for each activity.

1. 2 Styrofoam or paper cups of water [warm water differing in temperature by about 3°C (37°F)] and a thermometer

2. Illustration of lines in the text or worksheet

3. Washer on a string about 50 cm long, stopwatch

4. Illustrations in text or pictures of real plants, if available

5. 2 bottles filled about ¼ full of water: 1 bottle contains 1 drop of liquid soap, the other contains 2 drops (When the bottles are shaken, the mixtures should behave differently.)

6. 60 W and 40 W lightbulbs in holders (Hide the wattage label by coloring over the number with a marker.), pieces of paper

Suggested Approach

1. The six mini-activities in "You Be the Judge" are short, and it should be possible for most groups of students to carry out at least four activities in a single period. This is sufficient.

   - Some teachers choose to arrange the activities in the form of a circus and set up stations around the classroom.

   - Some teachers choose to have the apparatus available centrally. After completing an activity, students return the apparatus and collect whatever is required for the next activity.

   - Some teachers choose to do one or two activities with the class as a whole as a warm-up.

   The worksheet at the back of this guide (Appendix B, p. 64) provides a useful record and guide for the students.

2. Begin the discussion of the students' results by asking them what surprised them most when they were carrying out the activities. Use the section "Getting Help From the Scientists" to elaborate on and underscore students' feelings.

3. If time allows, discuss Philip's or Peter's investigation.

4. Assign "Back to the Project" as a homework assignment.

4

4** ARE SOME MEASURES BETTER THAN OTHERS?

Purpose

The lesson reviews the importance of quantifying (see previous lesson). It then proceeds to help the students understand and appreciate ideas about the reliability and sensitivity of different tests (means of measuring).

Time Allocation

Two 40-minute periods

Apparatus and Materials

Each pair of students will require a pair of magnets to test. In addition, there needs to be an accessible supply of the following materials that students can use to test the magnets:

- Paper clips
- Plotting compasses
- Pins
- Nails
- String or thread
- Copper wire
- Washers
- Iron filings
- Styrofoam blocks
- Rulers
- Rubber bands or elastics
- Plastic squares or squares made from note cards
- Spring balances; modeling clay (optional)

Suggested Approach

Period 1

1. Introduce the task of finding out which magnet is stronger. You might simulate Marie and Monique's project to provide a real-life context.

2. Read the section Help for Marie and Monique. This provides some guidance to the students as to how they might approach the task.

3. Give each pair of students 2 magnets and have them work with the magnets for the remainder of the period.

4. Assign the letter-writing task as homework. This is designed to help students reflect on their experience.

Period 2

5. Collect some of the students' answers to the problem. List them on the blackboard.

6. To launch a discussion that will likely raise the issues about quantitative (versus qualitative) observations, reliability, and sensitivity, invite 4 or 5 pairs of students to present their "best tests." (The overhead projector can often be used to silhouette their demonstrations.) Arrange for each pair to use the same two magnets (A and B) so that their results can be compared. (List the results on the blackboard.)

7. Role-play Marie and Monique's conversation with the scientist, pausing from time to time to relate the scientist's comments to the

previous discussion. This leads to the task at the end of the conversation.

8. There are many ways to handle the task of exploring the scientist's suggestions:

   • They might form the basis of a lab class.

   • The teacher might demonstrate them.

   • Volunteers might be asked to carry them out at home and report back.

Note: This lesson is based on a research study (Carlisle 1989) of 12 students in grade 7 who carried out a similar task. Here are some highlights from that study:

1. The three most popular types of tests were as follows:

   • The stronger magnet will hold more.

   • The stronger magnet will attract from a greater distance.

   • The stronger magnet will penetrate a barrier more easily.

2. Students collected data both qualitatively and quantitatively.

3. Most students standardized (controlled) the procedure they used within an individual experiment.

4. Each pair conducted between 2 and 15 experiments.

5. Individual experiments often show conflicting results. Only one pair correctly identified the strongest of 3 similar magnets, although 81% were confident they were right: The strongest magnet was the one that "won" most often.

6. Only one pair recorded their results.

4*** DESIGNING YOUR OWN MEASURES

Purpose

In devising their own measures and measuring instruments, students need to give thought to standards and units of measurement. After they understand these concepts, ideas of a scale and calibration are introduced.

Time Allocation

Four 40-minute periods

Apparatus and Materials

• Supply of plastic cups (about 150)

• Vials (about 15 [1 per pair of students])

• Medicine droppers (about 15 [1 per pair of students])

• About 3 L of 3 different brands of tea (These can be prepared before class.)

Safety Note

In general, students shouldn't eat or drink in the laboratory, so you will need to discuss this exception with them.

Suggested Approach

1. Introduce the "Desert Island Scenario." The cartoon is designed to prompt the students to think about the need to measure. The students might record their answers in three columns:

   - What to Measure

   - Units of Measurement

   - Instrument

   Before dividing the students into groups, you might provide one or two examples.

2. Review some of their answers. Elicit their answers through questions about the idea of the need for standards.

3. Read the section following "Desert Island Scenario." The students might be interested in knowing how the kilogram and a second are defined. A kilogram is the mass of a cubic decimeter of pure water (at the temperature of its maximum density). A second was formerly defined in terms of the time it takes for the Earth to circle the Sun; now it is defined in terms of a cesium-beam atomic clock.

4. Brew three of four samples of tea for the students to test. Allow them to explore the problem in groups.

5. Review the "Questions for Discussion." The first question invites discussion of the need to measure (quantitative versus qualitative). The second question makes a good demonstration. The colors show up well on an overhead projector.

6. The piece about "Measuring Instruments and Units of Measurement" is designed to be interactive. The questions in italics make for good discussion. If students have difficulty with the concepts of standards, units, and calibration, it may be necessary to introduce more examples for discussion. At the end, discussion returns to the question of how to measure the strength of tea.

7. The "Measurement Brainteasers" provide a review of notions of validity, quantification, sensitivity, reliability, and calibration. At the same time, the brainteasers also challenge one's ingenuity. You might ask students to choose one and work on it in small groups. Alternatively (and probably better, since they can experiment with it at home), you might invite them to do work on one as a homework assignment. The next section, "Advice From a Scientist," discusses the nature of each problem and provides clues on how it might be solved.

8. Review the students' solutions and discuss each in the light of "Advice From a Scientist." Take the opportunity to demonstrate some of the other solutions people have devised.

   a. Students might find it worthwhile to compare a biodegradable garbage bag with one that has been exposed to sunlight for several months. The original investigators used a strength test to tackle this problem.

b. As iron rusts, it combines with oxygen from the air. This happens rapidly with iron wool. The instrument (see diagram) is so sensitive that it can be used to find out which rusts faster, fine or coarse iron wool.

Iron wool

Water

c. A student who was measuring speed of rotation chose to measure the time it took for the axle of the windmill to wind up a 5 m length of thread.

d. A number of ingenious devices have been designed to measure air currents. One example is shown in the diagram.

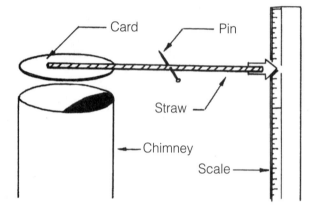

Card   Pin

Straw

Chimney

Scale

Chapter 5:
Variables and Their Controls

Middle school students seem to have an intuitive grasp of how to proceed with consumer or product testing. They frequently choose projects such as "Which detergent removes stains best?" or "Which battery lasts longest?" or "Can people really tell the difference between Coke and Pepsi?" In carrying out projects like these, students seem to give attention to the idea of being fair.

This chapter builds on the idea of the fair test and introduces the notion of controlled experimental conditions. In the second lesson, Finding Out How Much It Matters, attention is focused on the systematic variation of one variable. We will also consider the problem of identifying and separating variables in complex situations.

The experiment is central to much of scientific inquiry. Through controlled experiments, scientists try to develop their understanding of a phenomenon by ascertaining which factors affect the phenomenon and the extent to which each factor makes a difference.

Here is an extract from the *National Science Education Standards* (NRC 1996):

Design and conduct a scientific investigation. Students should develop general abilities, such as systematic observation, making accurate measurements, and identifying and controlling variables. (p. 145)

Here is an extract from *A Framework for K–12 Science Education* (NRC 2012):

Planning and designing such investigations require the ability to design experimental and observational inquiries that are appropriate to answering the question being asked or testing a hypothesis that has been formed. This process begins by identifying the relevant variables and considering how they might be observed, measured, and controlled (constrained by the experimental design to take particular values). (p. 59)

Chapter 5

5* BEING FAIR

Purpose

From an early age, children have a good idea of what being fair means. In a fair test, everything is treated in the same way (controlled), apart from that which is deliberately changed or varied. In this way, scientists can determine whether a change makes a difference.

Time Allocation

One 40-minute period

Apparatus and Materials

None required

Suggested Approach

1. The cartoon is designed to interest the students and help them develop a critical frame of mind. It should help them formulate questions (in small groups) about the experiments with the radishes and the glue.

2. Discuss the sections "Being Fair" and "Redi's Experiment."

3. Read about how the girls tried to control their hairspray experiment ("A Fair Test?"), and invite the students to respond to the questions at the end.

4. Ask the students to design fair tests for one of the projects described at the end of the lesson (a valuable homework assignment).

   Note: If students are having difficulty with the first part of the lesson, it might be worth extending it, using the reaction time experiment ("Sampling: An Introduction") described in Chapter 7 of the book for grades 8–12. Give students the task of formulating the rules for a reaction time competition. In doing so, they will identify variables that might make a difference in the results.

5** FINDING OUT HOW MUCH IT MATTERS

Purpose

This lesson builds on the idea of a fair test (5*). After scientists have learned (through a controlled experiment) whether or not a variable makes a difference, they often proceed to find out how much difference it makes. This involves the design of a series of controlled experiments in which one factor is systematically varied.

Time Allocation

Two 40-minute periods

Apparatus and Materials

Each group of students (3 or 4 per group would be optimal) will need the following:

- About 4 ft. (2 lengths) of Hot Wheels track (affordable and found in many toy stores)

- Ball about 3 cm in diameter (e.g., a table tennis ball)

- Ruler

- Protractor

Suggested Approach

1. Introduce the task and briefly discuss the sort of procedures the students might use.

2. Ask them to complete the questions for discussion (possibly for homework).

3. Review some of their findings.

4. Invite the students to do the "Quick Quiz" and read the section "Scientists Want to Know How Much?" When reviewing their answers, you might like to demonstrate the effects of varying the mass of the pendulum bob and the cooling of mugs containing different amounts of water (use cans to get a more dramatic effect).

5. Finally, invite the students to respond to the problems from "Back to the Project."

5*** ISOLATING VARIABLES: REDUCING COMPLEXITY

Purpose

Many scientific, technological, and engineering problems are extremely complex. This is especially evident in such fields as meteorology and engineering. When scientists encounter complex problems, they often use a strategy of reductionism, which reduces complexity by studying the effects of different variables in isolation. This lesson introduces students to this strategy.

Time Allocation

Two 40-minute periods

Apparatus and Materials

None required (although the apparatus described in the case study may be worth demonstrating and using with some simple shapes)

Suggested Approach

1. Introduce the case study and set the task as a small-group assignment.

2. Review and discuss answers from two or three groups.

3. The section "How Scientists Approach Challenging Problems" is designed for interactive reading and discussion. The middle section, which involves designing experiments with dowel rods, may warrant individual or small-group work.

4. The problems in "Back to the Project" may be used to stimulate class discussion or as homework assignments.

Notes

a. At least 10 variables affect the speed of boats. In addition to those mentioned, students might suggest the following variables: the angle of the bow to the water, the surface area of the hull, and the width of the boat.

b. Most of the variables are connected. Some may be combined if surface area is considered to be just one variable, but this subtlety is probably best left unmentioned unless it is raised by the students. At first sight, the salinity of the water is a separate variable, but remember that boats ride higher in salty water. The finish of the boat is a separate variable.

c. The dowel rod offers the opportunity to study the effects of area of cross-section, length, and weight, though weight is a difficult variable to isolate.

d. Square or rectangular blocks provide a simple means of studying the effects of varying how much surface area is submerged. You can vary the weight by adding cargo. The use of the balls begins to address the problems of shape of bow and stem. Some interesting possibilities emerge if sections of the spheres are stuck onto the dowel rod.

Chapter 6:
Experiment Design

Many scientists can recite horror stories about how the quest for reproducibility has caused problems in their experiments. They recognize that their findings are worthless without reproducibility.

Students know this intuitively; they expect things to happen the same way again and again. For them, this is what distinguishes the real world from magic. Yet they seldom use the idea of reproducibility (the word *repeatability* is used here) to find out if their experiments are working well—that is, if their experiments are well controlled. The first lesson is designed to help them see that repeatability is the test of a fair test.

In practice, scientists often seek experimental control (reproducibility) through the use of carefully designed apparatus. Indeed, apparatus is often designed with reproducibility in mind. The second lesson attempts to teach this idea.

One of the reasons scientists depersonalize descriptions of their procedures may be the desire to indicate that the procedures are so well controlled that they will work for anybody. This idea is central to the third lesson, which compares science with cookery.

Here is an extract from *A Framework for K–12 Science Education* (NRC 2012):

> As they become more sophisticated, students also should have opportunities not only to identify questions to be researched but also to decide what data are to be gathered, what variables should be controlled, what tools or instruments are needed to gather and record data in an appropriate format, and eventually to consider how to incorporate measurement error in analyzing data. (p. 61)

Chapter 6

6* GETTING EXPERIMENTS TO WORK: REPEATABILITY

Purpose

Students often assume that their tests are fair and their experiments are well controlled. Indeed, they seldom test for reproducibility (or repeatability, as it's called here). This lesson introduces students to the idea of repeatability—the test of a fair test. Repeatability is an important cornerstone of scientific exploration. If phenomena weren't repeatable, if things behaved in an irregular, unpredictable way, then science would not exist!

Time Allocation

Two 40-minute periods

Apparatus and Materials

- 2 or 3 raw eggs and 1 hard-boiled egg (You should have a few spare eggs on hand in case of breakage.)
- 2 beakers (about 500 ml capacity)
- Table salt
- Stirrer or spoon

Suggested Approach

1. Ask the students to complete "Quick Quiz: Predictions Please!" The quiz invites them to make explicit their understanding that we expect things to happen in the same way if an experiment is repeated.

2. Discuss their answers.

3. Read "The Test of a Fair Test" and relate it to the previous discussion.

4. At this point, you might introduce an activity that is not referred to in the text, "Egg Spinning Competition." Set up the competition on a table in front of the class. Invite the students to compete in groups of three or four. The winner goes to the playoffs! The students are free to select their egg, but unknown to them, one is hard-boiled. This one spins much better. After a few rounds, the students will likely notice that one egg spins differently. At this point, you might naively ask why they think this is the case. After confessing, you might ask what led them to think they were being tricked. This leads to the idea of repeatability: If identical eggs are spun exactly the same way, then they will all spin for the same amount of time.

5. Show the students the demonstration "Egg Magic" and invite them to find out if young people have developed the idea of repeatability.

6. Read the sections "Magic and Science Don't Mix" and "Checking for Repeatability." Pause to discuss the two problems from the science fair projects.

6** DESIGNING GOOD APPARATUS

Purpose

This lesson draws on (and reviews) a number of ideas from previous lessons:

a. Clarifying the meaning of the "best" bubble solution (4*, "Learning to Play the Numbers Game")

b. The idea of a fair test or controlled experiment (5*, "Being Fair")

c. The idea of repeatability (6*, "Getting Experiments to Work: Repeatability")

The lesson then invites the students to use these ideas creatively in the design of apparatus. If time allows, students can build on Lesson 5**, "Finding Out How Much it Matters."

Time Allocation

Two 40-minute periods

Apparatus and Materials

- 2 L bubble solution (see recipe in text or use 5% Ivory liquid soap). Divide in two and color one half with red food coloring and the other with green.

- 50 Styrofoam cups (reusable)

- 30 pieces of covered iron wire (25 cm lengths)

- Plastic drinking straws (about 30)

- Bicycle pump and small fish-tank aerator (if possible)

Suggested Approach

1. Show the students the two bubble solutions. (You can use the same solution, coloring one batch with green food coloring and the other with red.) Their task is to decide which solution is better for bubble blowing. Collect ideas from the students about how they might proceed. (Keep your comments to a minimum.) You might need to introduce a couple of ideas for bubble blowers (iron wire circles, drinking straws). Set them to work.

2. Field tests indicate that students find blowing bubbles very involving but tend to forget the task in favor of trying to make the biggest bubble. After about 20 minutes, you might organize a discussion, restate the task, and ask again for suggestions about procedure.

3. After a period of further experimenting, review some of their experiments, using the questions for discussion. Begin filling in the table.

4. Read the section "What Makes a Good Experiment? A Scientist's View" and link it to the students' own attempts through discussion.

5. Read and discuss the section "Designing Apparatus" with students and ask them to create additional designs.

6. Ask students if they think a bicycle pump would provide a good way to control air flow. This works well and is worth demonstrating. An even better source of controlled air supply is a small fish tank aerator. This

produces beautiful bubbles when fitted with a drinking straw.

7. If time allows, you might invite students to outline a range of tests that try to establish whether the bubble solution recipe given in the text can be improved.

8. Ask the students to tackle one of the challenges and devise apparatus that can be used in well-controlled tests.

6*** PREPARING EXPERIMENTAL DESIGNS

Purpose

The problem of turning cooking (in this case, making cookies) into a science has many facets to it:

- It involves the reformulation of a recipe to add control and express it with precision in scientific terms.

- It involves the identification of dependent and independent variables. (Dependent variables such as lightness and crunchiness need clarification, and their measurement requires considerable ingenuity.)

- It involves preparing a series of experiments—an overall design—that enable the problem to be investigated systematically. The lesson reviews the two previous lessons, "Designing Your Own Measures" (4***) and "Isolating Variables" (5***). It then involves the students in the problem of preparing experimental designs.

Time Allocation

Home assignment (optional), plus one or two 40-minute periods

Apparatus and Materials

None required. Before the lesson, the students might be invited to prepare for the cookie contest.

Safety Note

In general, students shouldn't eat or drink in the laboratory, so you will need to discuss this exception with them.

Suggested Approach

1. Set up the cookie contest as a homework assignment. The recipe is provided in the text. This contest would enhance the lesson but is not crucial.

2. Discuss the contest using the section "Questions to Consider."

3. In small groups, give the students the assignment of devising a plan to improve the recipe.

4. Read "Turning Cooking Into a Science" and discuss the questions at the end of each paragraph.

5. Invite the students to plan procedures for investigating the two sample science fair projects or invite students to report on the ways in which they identified variables in their projects.

Chapter 7:
Sources of Error

If you look up the word *science* in a dictionary, you will probably find it defined using words such as *knowledge*, *exact*, and *accurate*. In short, science is about "knowing for sure." This chapter is about being accurate and being sure. It is about errors—recognizing them, taking account of them, and eliminating them. The chapter is designed to help students become aware of and respond to errors of measurement and errors that arise from a failure to control variables.

The first lesson focuses on errors of measurement. It introduces the ideas of taking the average and calculating the average error. (Many middle school students are aware of the idea of taking the average. For them, it seems there is a "magic" number of measurements to take.)

The second lesson examines the problem of tracking down errors arising from uncontrolled variables. It involves the students in a search for those factors that determine the reproducibility of their data. Once identified, it may be possible to control these factors, thus eliminating the errors.

There are, of course, some variables that are not possible to control. Problems of this sort are often encountered by scientists working with humans, animals, or plants. This lesson extends the idea of taking averages and sets the stage for an introduction to statistics and sampling.

An extract from *A Framework for K–12 Science Education* (NRC 2012):

> As in other forms of inquiry, the key issue is one of precision—the goal is to measure the variable as accurately as possible and reduce sources of error. The investigator must therefore decide what constitutes a sufficient level of precision and what techniques can be used to reduce both random and systematic error. (pp. 59–60)

Chapter 7

7* TAKING THE AVERAGE

Purpose

This lesson builds on the ideas of a fair test, a controlled experiment (Lesson 5*), and the test of a fair test—repeatability (Lesson 6*). Since perfect repeatability and reproducibility are seldom attainable, taking the average of a number of readings is often used to enhance reliability by ironing out errors of measurement and differences due to uncontrolled variables. This lesson introduces students to different sources of error and to the idea of averaging.

Time Allocation

One or two 40-minute periods

Apparatus and Materials

- Meterstick
- Table tennis ball

Suggested Approach

1. Read the cartoon and then have five members of the class help read the height of bounce of a table tennis ball when it is dropped from 1 m high. Record the readings on the board.

2. Read the section "The Best Answer." Involve the class in discussing the questions in italics.

3. The section "Two More Problems" contains more questions for discussion. Alternatively, the questions may be given as a homework assignment.

4. When possible, students should apply these ideas to their own project (last section).

7** SEARCHING FOR ERRORS

Purpose

Scientists often encounter the problem of lack of reproducibility; their results fluctuate so much that they are unable to proceed. This is frequently due to the effects of uncontrolled variables. They often spend frustrating hours trying to identify and then control these variables to obtain results that are reproducible. This lesson has students searching for errors in the form of uncontrolled variables.

Time Allocation

One or two 40-minute periods

Apparatus and Materials

Each pair of students will need a jar or beaker (about 500 ml) and 1 or 2 candles.

Suggested Approach

1. Refer to Starting Point 10, "Suffocating Candles," and demonstrate the extinction of a candle flame using a medium-size jar (about 500 ml). Time how long the candle stays alight. Repeat the experiment, and time again. The two times are often very different. Repeat again if necessary.

2. Invite a brief discussion about why the times are different and then read the section "Wayne and Darrell Need Help."

3. Challenge the class to get the experiment to work.

4. Discuss some of the students' findings: "What did you think the problem was?"

"Why?" "How did you try to remedy it?" "How did you show your remedy worked?" Read the section Identifying Uncontrolled Variables, pausing for comment or discussion.

5. Demonstrate the importance of flushing out the jar with fresh air. Between each experiment, wave the jar back and forth half a dozen times. This process yields fairly reproducible results.

6. Tell your students that their observations of their experiments should say how the conditions were controlled. For example, Wayne and Darrell should say something like "The jar was flushed with fresh air between experiments by waving it back and forth six times."

7*** SAMPLING: AN INTRODUCTION

Purpose

This lesson builds on the idea of averaging to iron out errors (6*). It invites the students to consider how many readings should be taken so that they can be reasonably sure their findings are significant. It provides an elementary introduction to sampling and probability.

Time Allocation

Two 40-minute periods

Apparatus and Materials

10 rulers

Suggested Approach

1. Catch the students' interest by challenging one or two to catch a $5 bill.

2. Show them how they can measure reaction times using a ruler.

3. Divide the class into groups of three or four: two or three competitors, a judge, and a scorer. Ask the groups to make 10 readings for each competitor.

4. Read about Adrian's and Matthew's experiments. Pause at the end of each paragraph to discuss the question asked.

5. Invite the students to work out their averages and their running averages (average after each new reading).

6. Optional: If the students show a strong grasp of these ideas, you might introduce the idea of sampling. To do this, you might compare the average performance of boys and girls in the class, then ask if it would be fair to extend any conclusion to all classes in the school, to all schools in the state, to all states in the country, and finally to all countries in the world.

7. In small groups, or for a homework assignment, ask the students to comment on the results obtained by two grade 8 students. Ask for suggestions as to how the experiments could be improved:

   a. Should they have taken more readings?

   b. Should the student examining the effects of salt concentration have made measurements at more concentrations (to smooth out the curve obtained)?

   c. Should the student examining flight have tried to design a controlled way to launch his airplanes?

   d. Is it fair for these experimenters to generalize across all plant crops or all paper airplane throwers?

Note: The reaction time experiment can also be used to introduce the idea of variables. Students can be asked to formulate the rules for a reaction time competition. In their desire to be fair, they will of course identify possible variables.

One commonly forgotten variable is practice. Matthew's performance suggests that he may improve with practice.

These ideas might be worth weaving into the lesson if the opportunity presents itself.

Chapter 8:
Making Sense of Your Results

This chapter is about graphing data. The three lessons are designed to develop the students' awareness of the value and techniques of graphing; they provide an overview of graphing rather than a comprehensive course in graphing skills.

The first lesson simply attempts to heighten students' appreciation of the value of picturing their results graphically.

The second lesson invites them to consider some of the decisions they need to make when plotting graphs, such as about the type of graph, the direction of the axes, scale, and joining of points.

The third lesson portrays graphing as a picture language that complements the language of numbers and facilitates the interpretation of numerical data. The students are informed about a variety of common graph shapes and are tacitly introduced to ideas about slopes, extrapolation, interpolation, and intercepts.

Here is an extract from the *National Science Education Standards* (NRC 1996):

Use appropriate tools and techniques to gather, analyze, and interpret data. The use of tools and techniques, including mathematics, will be guided by the question asked and the investigations the students design. (p. 145)

Here are extracts from *A Framework for K–12 Science Education* (NRC 2012):

Once collected, data must be presented in a form that can reveal any patterns and relationships and that allows results to be communicated to others. Because raw data as such have little meaning, a major practice of scientists is to organize and interpret data through tabulating, graphing or statistical analysis. (p. 61)

In middle school, students should have opportunities to learn standard techniques for displaying, analyzing, and interpreting data; such techniques include different types of graphs, the identification of outliers in the data set, and averaging to reduce the effects of measurement error. (p. 63)

Chapter 8

8* CHARTING YOUR DATA

Purpose

This lesson is designed to help students appreciate the value of charting their data in the form of bar graphs.

Time Allocation

Two 40-minute periods

Apparatus and Materials

- String (about 50 m)
- Washers (1 per student)
- Masking tape (about 2 m)
- Large clock with second hand (if students don't have watches)

Suggested Approach

1. Give each pair of students 1 m of string and a washer; set the challenge of making a 1-minute timer.

2. Review students' answers and ask a few students to describe how they did the task.

3. Read the section "How Did You Solve the Problem?"

4. Give each student a length of string and a washer. Make an "instant picture" as a whole-class activity. (Ensure that there is wide variation of string length around the class. You call out "start" and "stop" at the beginning and end of the 10-second period.)

5. Discuss the picture produced ("Something to Think About").

6. Assign the exercise "Making a Bar Graph" (possibly as a homework assignment).

8** GRAPHING YOUR DATA

Purpose

This lesson provides a rapid overview of some of the important decisions made in graphing data: decisions about the type of graph, the orientation and scale of the axis, and the joining of points.

Time Allocation

Two 40-minute periods

Apparatus and Materials

None required

Suggested Approach

1. Introduce the students to the data showing how high a table tennis ball bounces when dropped from various heights.

2. The data can be plotted in many ways. The students' task (in groups of two) is to make five decisions about the alternatives

presented. Encourage them to formulate their answers to the questions in writing.

3. The section "Kim and Lee Talk to a Scientist" provides insights into which alternative is best. Review the students' answers in a question-and-answer discussion, then help

them compare their answers with those given by the scientist.

4. The sections "Questions for Discussion" and "Advice for Al and Art" provide an opportunity for students to contextualize and consolidate the ideas they have met (possible homework assignment).

8*** INTERPRETING GRAPHS

Purpose

The major purpose of this lesson is to help the students appreciate the value of graphs—in particular, that graphs are more informative than a string of numbers. It provides a brief overview of some important ideas: slopes of graphs, intercepts, extrapolation, and interpolation.

Time Allocation

Two 40-minute periods (*Note:* If desired, introduce the lesson at the end of the previous class and distribute beans for students to grow at home.)

Apparatus and Materials

- 1 or 2 bean seeds per student
- Toy car

Suggested Approach

1. This lesson begins with a case study on the growth of a bean. To enhance the lesson's meaning, give the students a bean to grow one week before you intend to start the lesson.

2. At the beginning of the lesson, ask the

students about their observations: "What did you observe?" "How much did yours grow?" "When was it growing fastest?" The last question cannot, of course, be answered unless students have collected some data.

3. Invite the students to plot Sam and Sid's data.

4. Read the section "Number Languages, Picture Languages, and Words," asking questions that help the students to relate the topic to their studies or to Sam and Sid's study of bean growth.

5. Acting as Interpreter: Invite the students to work in pairs or small groups to interpret the graphs of velocity against time. Tell them they should be prepared to demonstrate their answers with a toy car on the teacher's desk.

6. Review their answers in a whole-class discussion:

   a. The car is moving at a constant speed of 30 m/s.

   b. The car is accelerating. Its maximum acceleration occurs at about 5 seconds, when it reaches a top speed of 50 m/s. The graph is much easier to read than the numbers.

c. Both drivers take 4 seconds to stop, but driver A has better brakes. The slope of the stopping curve for driver A is steeper.

d. The engine might have failed and the car may be coasting to a stop. If you extrapolate the curve, you can estimate that the car stops in about 2 more seconds (at time = 12 seconds).

e. The car was moving at 10 m/s when the clock started (the value of the intercept on the velocity axis). If you extrapolate (backward), you can estimate that the car started 2 seconds before the clock.

f. The car was probably moving at 40 m/s when its engine failed (at about time = 4 seconds). It will probably stop at about time = 14 seconds.

7. Ask the students to do one of the examples from "Back to the Project" as a homework assignment. Here are the answers:

1. When dropped from about 10 m, the ball has reached its terminal velocity as it hits the ground: It can't go any faster because air resistance is equal to the force of gravity. The height of bounce depends on the ball's velocity at impact and therefore levels out.

2. The fitness of an athlete can be measured by his or her recovery time—the quicker the recovery, the fitter the athlete. There are two problems to be solved now:

a. What do you do when two athletes have different resting pulse rates?

b. Can you find a simple way to measure the recovery time from the graph?

There are many possible solutions to these problems.

Chapter 9:

Explanations

This chapter focuses on the products of scientific inquiry. It could easily be considered the most important chapter in this book because it relates the product of inquiry to the process of inquiry and presents scientific activity as a human endeavor motivated by the desire to produce valuable products (products that are useful in a broad sense).

Most would agree that the general purpose of scientific inquiry is to develop a comprehensive understanding of the world in which we live. However, it is not easy to say exactly what is the nature of that understanding—the product of scientific inquiry. Some portray scientific inquiry as a set of concepts, some as theories, and some as laws that govern the behavior of the natural world.

This curriculum portrays this understanding as a network of explanations about the world. Three sorts of explanations are considered: interrelated patterns of behavior, trends in relationships, and analogies and models. Here is a brief description of each explanation:

a. *Explanations as interrelated patterns of behavior:* The product of scientific inquiry is more than a collection of useful facts about the world. Although it is useful to know that battery X lasts for 365 minutes and battery Y lasts for 308 minutes, it explains nothing.

Developing an explanation (or understanding) involves perceiving patterns among the facts and then relating one pattern to another in the form of a generalization—such as alkaline batteries last longer than general purpose batteries.

b. *Explanations as trends in relationships:* To say potatoes grow better in acid than in alkaline soils certainly links two patterns (growth and acidity) and therefore can be considered an explanation, but it is not generalized to extremes of acidity and alkalinity. Scientists try to find out more about how the growth of a plant is affected by acidity.

c. *Explanations as analogies and models:* The analogy has been made between the Earth's atmosphere and a blanket. The kinetic theory uses the particle model of matter. Scientists use analogies and models such as these to make further connections between observed patterns of behavior and add an extra dimension to their understanding. Good analogies and models connect many seemingly disparate findings and provide coherence to our understanding. Their power rests in their capacity to explain diverse observed phenomena and predict that not yet observed. Scientists have found that successful and long-lasting analogies and models (theories) are characterized by their simplicity. In sum, a good analogy or model provides coherence, has explanatory and predictive power, and is simple.

Chapter 9

The first lesson invites the students to express the findings from their experiments in the form of generalizations (interrelated patterns and/or trends). At the same time, it emphasizes the value of these findings.

The second lesson is concerned with getting explanations that fit and are coherent. The lesson portrays scientific inquiry as a search for that coherence through a process of experiment, explanation, prediction, further experiment, and so on.

The third lesson builds on the first two lessons and introduces the possibility of deepening our understanding through analogies and models.

Here is an extract from the *National Science Education Standards* (NRC 1996):

> Develop descriptions, explanations, predictions, and models using evidence. Students should base their explanation on what they observed, and as they develop cognitive skills, they should be able to differentiate explanation from description—providing causes for effects and establishing relationships based on evidence and logical argument. (p. 145)

Here are extracts from *A Framework for K–12 Science Education* (NRC 2012):

> Scientific explanations are accounts that link scientific theory with specific observations or phenomena—for example, they explain observed relationships between variables and describe the mechanisms that support cause and effect inferences about them. (p. 67)

> Because scientists achieve their own understanding by building theories and theory-based explanations with the aid of models and representations and by drawing on data and evidence, students should also develop some facility in constructing model- or evidence-based explanations. (p. 68)

9* LOOKING FOR PATTERNS AND TRENDS

Purpose

This lesson is designed to introduce the students to the idea that the product of scientific inquiry is more than a collection of experimental facts. It is the connections that link them.

The lesson introduces the students to two types of connections: connections that link patterns (or categories) of behavior and connections that describe in some detail the way the patterns are related (i.e., how one varies with another). The lesson introduces the term *generalization* to describe these connections.

Finally, the students are invited to consider the value of generalizations. They are often useful, of course, but more than that, they prompt scientists to inquire further into the regularity of natural phenomena. This sets the stage for the next lessons.

Time Allocation

Two or three 40-minute periods

Apparatus and Materials

Wood and nail samples

Suggested Approach

1. Present Ravi's findings on the holding power of nails in wood. Ask small groups of students to write up the conclusions. Encourage them to look for patterns and trends, such as that hard woods hold more firmly than soft woods, brittle materials don't hold as well as wood, and dry and old wood holds more firmly than wet and new wood. You might need samples of these materials and the nails to help the students in their search for significant qualities (patterns).

2. Read the sections "Looking for Patterns" and "Looking for Trends," pausing at the end of each paragraph to discuss the italicized question.

3. Read "Robert Hooke's Claim to Fame" and ask the question at the end of the piece (whole-class, small-group, or individual work).

4. "Back to the Project" may be given as class work or as a homework assignment.

Chapter 9

9** GETTING EXPLANATIONS THAT FIT

Purpose

This is probably one of the most important lessons in this book. It is designed to help students conceptualize some major aspects of both the product of scientific inquiry (scientific explanations) and the process by which these are progressively developed. It reveals the interplay between explanation and experiment. This interplay of "conjecture and refutation," as Popper calls it, is presented through teacher demonstration as one that often follows the sequence: experiment, explanation of findings, prediction, further experiments, and confirmation or revision of explanation.

Time Allocation

Five 40-minute periods (It's a pity to rush this!)

Apparatus and Materials

It's sufficient for the teacher to demonstrate the experiments in this lesson. The following activities are required:

- "The Amazing Can": See diagram on the right in part 1, step 2.

- "Prediction Challenge": retort stand, 2 clamps, 2 bosses, string, and weight

- "Strange Happenings": funnel, table tennis ball, 2 apples, string, retort stand, clamp, and boss

- "Confirming Explanations": candle and beaker

Suggested Approach

Part 1: Fitting New Findings in With Existing Explanations

1. Introduce Wayne and Darrell's results and ask the students to write down their suggested conclusions. Discuss some of these and keep some for discussion at the end of the lesson.

2. Introduce "The Amazing Can" by reading the section "Scientific Explanations." Then demonstrate the activity. Ask the students to explain, with the aid of a diagram, how it works. If time allows, you might ask the students to suggest how they might test their explanations (without opening the can).

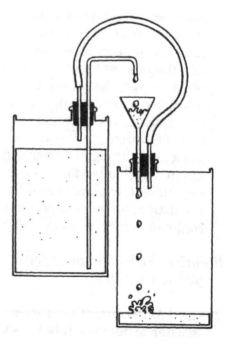

Part 2: Using Explanations to Predict the Unknown

3. Set up the Prediction Challenge. Here the students explain their answers before they carry out the test. Review their answers and then discuss how they went about tackling the problem.

   John O'Connor (1990) wrote in the *SI²
   Network Newsletter*, "The bob tries to circle the bar. To complete the loop the bob must be released from a point twice the radius of the circle the bob describes around the crossbar. ...This activity is for me the highlight of the year. … My problem is that this process takes three periods, not half of one! Am I overestimating the difficulties that grade 8 students experience with energy concepts?" (*Note:* SI² stands for "Student's Intuitions, Science Instruction.")

Part 3: Modifying and Confirming Explanations

4. Invite the students to predict the outcome of "Strange Happenings" and then demonstrate the effects. Read the section "Modifying Explanations" and immediately assign the problems of the bending candle flames ("Confirming Explanations").

5. Finally, read the sections "Scientific Discoveries" and "Have You Found Anything New?"

6. For homework, invite the students to choose an assignment from "Back to Your Projects."

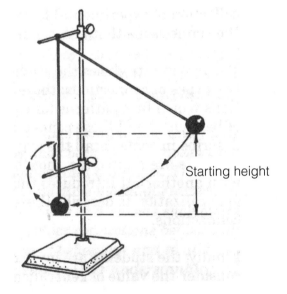

Starting height

Chapter 9

9*** DEEPENING YOUR UNDERSTANDING: ANALOGIES AND MODELS

Purpose

The lesson is designed to help students begin to appreciate the way scientists attempt to deepen their understanding by proposing analogies and models of the regularities they find in nature. Analogies and models add an extra dimension to the regularities observed and often can be used to link these regularities together. The lesson is not an easy one for students to grasp, but some middle school students should be able to appreciate the value of analogies.

The lesson proceeds to examine what scientists consider the features of good models or theories. They not only help link a wide range of phenomena. They have a wide explanatory power and are coherent but also prompt prediction and further inquiry. Two other features of models are touched on: They are not "real," and good models are simple and elegant (Occam's razor).

Time Allocation

Three 40-minute periods

Apparatus and Materials

- Black box
- 3 knitting needles
- 3 washers
- 6 elastic bands

Note: You might need 2 or 3 sets of these materials.

Suggested Approach

1. Introduce Heidi's project and ask students (in their small groups, perhaps) to answer the three questions about which analogy they find the best.

2. Read the section "Scientists Use Analogies," pausing to discuss the questions at the end of each paragraph. This provides an opportunity for students to discuss the analogies they have encountered in the course of their science studies—for example, light bouncing like billiard balls or plant leaves acting like a food factory. Analogies have their limits, and the limits of those that compare wood to rubber or cloth are recognized. It might be useful to refine these analogies and ask the students to examine the way in which wood behaves like a roll of cloth or a bundle of elastics. It could prompt them to ask further questions, such as "Does wood have holes in it like cloth?" and "What sort of glue holds the elastics together?"

3. Read about scientific models again, pausing to discuss the questions at the end of each paragraph.

    The students might be interested to hear about how Aristotle's model was developed. His fellow astronomers were very interested in his ideas and began to plot the movement of the planets very carefully. They noticed that two of the planets, Venus and Mercury, which were never far from the Sun, sometimes seemed to be travelling faster and sometimes slower. How could this be explained? Aristotle's theory did not look so good! In the 16th century, 1800 years later, Copernicus thought of an alternative explanation: Perhaps the Sun is at the center of the universe. Perhaps Earth and the planets go in circles around the Sun. Perhaps the stars are fixed in the heavens. Perhaps Earth spins on its axis once a day. Perhaps the Moon goes around Earth. This was a wonderful explanation. It seemed to account for everything that was known. However, when scientists checked with special instruments the paths of the planets, they found that the data did not quite match.

About 50 years after Copernicus died, Kepler began to try to find a simple way to describe the paths of the planets that fitted the data. Perhaps the planets move around the Sun in an ellipse. Perfect fit! But why do the planets travel in ellipses? That was left for Newton to explain.

4. Demonstrate the "Black Box Puzzle" and ask the students to compare the puzzle with a scientific model.

Similarities	Differences
Models are inventions of the mind.	There really is something inside the black box.
You can test your ideas.	

A Note on Mathematical Models

Mathematical models have not been discussed in this lesson. This could be considered a major omission since these models attempt to describe and explain the regularities of nature, such as the ebb and flow of the population of species, the rate of chemical reactions, and the electronic structure of atoms. But such matters are often complex.

However, if you want to introduce the mathematics model, you might choose a simple system to illustrate the concept—for example, the rise and fall of a population of mealworms over time.

Some computer screensavers provide an excellent analogy to mathematical modeling. For example, check out the Microsoft Windows (XP Home Edition) screen saver. Every few seconds, the Windows logo moves from one place to another on the screen, seemingly randomly. But is it really random? To understand the logo's movement, we need to find the pattern. Our understanding would be deepened if we were able to write a program that reproduces the pattern. In science, mathematics is often used to describe such patterns. In this way, mathematics deepens our understanding of our observations.

You might be able to find (through Google) the original and popular Macintosh screensaver "Pyro," which shows fireworks exploding. The height of the trajectory and the number of stars produced varies. Maybe your students would welcome the challenge of finding the pattern and even trying to reproduce it!

Chapter 10:
Sharing Your Findings

This chapter is about sharing and communicating scientific findings. It is designed to help students who have done a project display it, write it up, and talk about it. The chapter attempts to put this exercise in the context of how the community of scientists shares discoveries through writing (in journals) and talking about them (at conferences).

The first lesson prepares students for displaying their project at a science fair. It tries to do so in a fairly non-directive way (no standard format is prescribed).

The second lesson develops the theme from the section in Chapter 9**, "Have You Found Anything New?" This theme attempts to show students how scientists use journals to communicate their findings and how these findings eventually end up in textbooks. The students are then introduced to the impersonal style that characterizes much scientific writing and the way scientific papers are structured. Nobel Laureate Peter Medawar, by the way, considered the scientific paper a fraud because its structure (introduction, previous work, method, results, and discussion) misrepresented the process of scientific thought. His point is acknowledged. Finally, it's worth noting that scientific writing has a persuasive element. When discussing and interpreting the results, scientists often present arguments in support of their conclusions and sometimes even suggest ways to test them!

In the third lesson, with a view to encouraging students to talk about their projects, the students are introduced to the role of scientific conferences in enabling the community of scientists to communicate.

Here is an extract from the *National Science Education Standards* (NRC 1996):

> Communicate scientific procedures and explanations. With practice, students should become competent at communicating experimental methods, following instructions, describing observations, summarizing the results of other groups, and telling other students about investigations and explanations. (p. 148)

Here is an extract from *A Framework for K–12 Science Education* (NRC 2012):

> Students should write accounts of their work, using journals to record observations, thoughts, ideas, and models. They should be encouraged to create diagrams and to represent data and observations with plots and tables, as well as with written text, in these journals. They should also begin to produce reports or posters that present their work to others. As students begin to read and write more texts, the particular genres of scientific text—a report of an investigation, an explanation with supporting argumentation, an experimental procedure— will need to be introduced and their purpose explored. Furthermore, students should have opportunities to engage in discussion about observations and explanations and to make oral

presentations of their results and conclusions as well as to engage in appropriate discourse with other students by asking questions and discussing issues raised in such presentations. (p. 77)

10* DISPLAYING YOUR PROJECT

Purpose

There are no standard formulas for designing a display. It's a creative art! However, the first part of the lesson is designed to help the students identify some of the important things they might take into consideration. The second part gives them practice in trying out their ideas. It's hoped that when they reflect on this experience, they will be better able to prepare the display of their own project.

Time Allocation

One or two 40-minute periods (plus homework)

Apparatus and Materials

None required. A model of a display board and a good supply of scrap paper (used computer paper or unused newsprint) could be useful. If you decide to simulate Rich's project, you will also need the apparatus and materials listed in the next lesson.

Suggested Approach

Part 1

1. Take a look at each of the projects illustrated in the book (see photos on pp. 56, 58, and 59). Invite the students to say what they like and don't like about them.

2. Ask the students individually to write in their notebooks a checklist of features of good design.

3. Read "Designing Your Display" and briefly discuss each of questions 1–10. Ask the students to compare their checklists with the questions.

Part 2 (optional)

4. Set the stage. The students are given the task of translating "Rich's Science Project Diary" into a display. Briefly review what Rich did. You might wish to simulate this.

5. Invite the students to discuss the questions 1–10 ("Designing Your Display") with each other in pairs.

6. Give out scrap paper and set the students to work in pairs. Designing a display is not an easy task, and they will find this out for themselves. They are often tempted to put pencil to paper right away. You might find it useful to encourage them to think about layout first (question 7).

7. It's unlikely they will finish the task. Five minutes or so before the end of the period, have a discussion about the difficulties students encountered.

8. Ask them to finish the outline of their displays so they can be shown and discussed at the next class.

10

10** WRITING YOUR REPORT

Purpose

This lesson is designed to provide a forum for discussing the style and structure of a good report. An attempt is made to avoid rigid prescription for two reasons:

1. It is important for report writing to be an opportunity for sense making. Prescribing the impersonal may negate this.

2. The structure adopted by most scientific papers does not truly represent the twists and turns of scientific inquiry.

Time Allocation

Two 40-minute periods

Apparatus and Materials

Some apparatus will be required if parts of the described project are to be simulated (demonstrated). The following items should be sufficient:

*   Piece of plywood, 1 m × 0.5 m (approx.)

*   2 different cans of soup

*   Tin of chocolate powder

*   2 empty coffee cans

Suggested Approach

1. Introduce the students to "Rich's Science Project Diary." This is a delightful project, and you can demonstrate parts of it to bring it alive. (The apparatus requirements are very simple.)

2. Ask students to write a report on "Rich's Science Project." (They could complete this for homework.)

3. After inviting 1 or 2 students to read their reports aloud, discuss "Scientists Write Their Reports."

4. Pairs of students might then exchange their reports and comment on them in light of how they might change the report to fit the way scientists write.

Chapter 10

10*** TALKING ABOUT YOUR PROJECT

Purpose

The immediate purpose of this lesson is to prepare students for their encounters with judges at science fairs. However, the value of students organizing their own conference should not be underestimated. Apart from making considerable demands on the confidence of the presenters, this experience provides the opportunity for the audience to sharpen their skills in commenting on and constructively criticizing scientific investigations. (See 11*** for the development of this idea.)

Time Allocation

One or more 40-minute periods

Apparatus and Materials

- Piece of plywood
- 2 cans of soup
- Can of chocolate powder
- 2 empty coffee cans

Suggested Approach

1. Read the introductory section "Scientists Share Their Findings."

2. Refer to "Rich's Science Project Diary." The students might appreciate parts of the diary being simulated to bring it alive. The apparatus requirements are simple: a piece of plywood, two cans of soup, a can of chocolate powder, and two empty coffee cans.

3. Discuss how Rich might best present the project.

4. Ask the students if they think Rich made a good job of the project. After warming them up to making comments and constructive criticisms in this way, give them (in pairs) the task of devising one question, one comment, and one suggestion each.

5. If appropriate, follow this activity with one or two classes in which students organize their own "can" conference.

Chapter 11:
Judging Projects

This sequence of lessons follows naturally from those in Chapter 10. The first lesson focuses on judging displays, the second on judging written reports, and the third on judging oral presentations.

The lessons may also be considered a review of what students have learned. To help them make judgments, the students are invited to reflect on the experience they have had in the course they have taken.

Judging is a sophisticated activity. The first lesson attempts to simplify this by helping students devise their own checklist of the qualities necessary for a good project. This process is designed to help them appreciate the types of forms used by judges at science fairs. In reality, however, making judgments is more artful than simply "plugging in" a checklist. When making judgments, experienced scientists often compare the inquiry undertaken with the way in which they (ideally) would have done it. The third lesson introduces this idea.

Throughout the chapter, the judging of science fair projects is put in the context of the way scientists judge each other's work.

Here is an extract from the *National Science Education Standards* (NRC 1996):

Recognize and analyze alternative explanations and predictions. Students should develop the ability to listen to and respect the explanations proposed by other students. They should remain open to and acknowledge different ideas and explanations, be able to accept the criticism of others, and consider alternative explanations. (p. 148)

Here are extracts from *A Framework for K–12 Science Education* (NRC 2012):

Scientists and engineers use evidence-based argumentation to make the case for their ideas, whether involving new theories or designs, novel ways of collecting data, or interpretations of evidence. They and their peers then attempt to identify weaknesses and limitations in the argument, with the ultimate goal of refining and improving the explanation or design. (p. 46)

Students should begin learning to critique by asking questions about their own findings and those of others. Later, they should be expected to identify possible weaknesses in either data or an argument and explain why their criticism is justified. (p. 74)

Chapter 11

11* CHECKING FOR QUALITY

Purpose

This lesson provides an opportunity to review what the students have learned and develop their appreciation of the features of good projects as well as introducing them to the judging procedure.

Time Allocation

One 40-minute period

Apparatus and Materials

None required

Suggested Approach

1. Review the skills of an experienced scientist, listing them on the board.

2. Invite the students (in small groups) to judge the two projects illustrated and list the qualities of a good project.

3. Compile on the board the qualities of a good project.

4. Refer to the judging forms in the appendix and invite the students to compare them with the two lists on the board.

11** SUGGESTING IMPROVEMENTS

Purpose

This lesson invites students to suggest improvements to an actual science fair project by reflecting on what they have learned previously about scientific inquiry.

Time Allocation

One (possibly two) 40-minute period

Apparatus and Materials

None required

Suggested Approach

1. Set the stage for the task by reading about the job of an editor of a scientific journal.

2. In preparation, the students then read Danny and Adam's account of their project.

3. Introduce the task of suggesting improvements (the first question under the subhead "Becoming an Editor"). It is not an easy task, and you will probably find it necessary to build the students' confidence in tackling it by referring to some of the experiences they have had in previous lessons—for example, "Do you think they controlled their experiment well?"

11

4. Invite the students (working in small groups) to suggest three improvements per group.

5. Conclude the lesson by reviewing some of their suggestions. Did they consider the following questions?

   a. What is the best way to measure plant growth?

   b. What is the best number of beans to grow in each cup?

c. What is the problem with some of the colors being solid while others are transparent?

6. Optional homework: Either ask the students to take the role of an editor and write a letter to Danny and Adam, or ask the students to make suggestions about improving the report. (Refer them to section "Scientists Write Their Reports" in Lesson 10**.)

11*** MAKING JUDGMENTS

Purpose

The process of judging is presented as one in which the merits of alternative research designs are explored. At a conference, questions from the audience often lead to exploration of these alternatives. This lesson prepares students for participating in a conference of their own.

Time Allocation

One 40-minute period

Apparatus and Materials

None required

Suggested Approach

1. Involve the students in a brief discussion about the problem of making judgments (e.g., you might discuss ice skating or gymnastic competitions). Turn the discussion to the problem of judging science projects and read the section "What Would You Have Done?"

2. Invite the students to tackle the task of redesigning one of the projects displayed in Chapter 11***.

3. Use the questions to review some of their answers.

4. Read the section "In the Judge's Shoes" and ask the students to tackle the task "You Be the Judge." This provides them with practice if you are planning the sort of conference described at the end of Lesson 10***.

Chapter 12:

Generating Ideas for Projects

Although this is the last chapter, some teachers may choose to use it first. This chapter is designed to help students generate ideas for an investigation of their own and serves as a launching pad for science fair and science project work.

The three lessons in this chapter progressively increase in difficulty and authenticity. The first lesson invites students to review some successful science fair topics and use these as a springboard for their own ideas. The second focuses on the world around us as a source of interesting possibilities. The third portrays the scientific literature as an authentic source of ideas.

Here is an extract from the *National Science Education Standards* (NRC 1996):

> Identify questions that can be answered through scientific investigations. Students should develop the ability to refine and refocus broad and ill-defined questions. An important aspect of this ability consists of the students'

ability to clarify questions and inquiries and direct them toward objects and phenomena that can be described, explained, or predicted by scientific investigations. (p. 145)

Here are extracts from *A Framework for K–12 Science Education* (NRC 2012):

> Questions are the engine that drive science and engineering.

> Science asks:

> - What exists and what happens?
> - Why does it happen?
> - How does one know? (p. 54)

> The experience of learning science and engineering should therefore develop students' ability to ask—and indeed, encourage them to ask—well-formulated questions that can be investigated empirically. (p. 55)

Chapter 12

12* IDEAS FROM PREVIOUS SCIENCE FAIRS

Purpose

This lesson is designed to help students who have limited experience with science fairs generate an idea for a project of their own. If some still experience difficulty, it may be worth following the lesson with 12**, "Ideas All Around You."

Time Allocation

One 40-minute period

Apparatus and Materials

None required

Suggested Approach

1. Ask the students (working in pairs) to select two or three topics that interest them from the list of project titles.

2. Use the example of Tracey and Shelley's conversation to show students how to explore the scope of a topic.

3. Further exploration (for homework) can follow using the ideas in Lesson 3*, "Wondering Why."

12** IDEAS ALL AROUND YOU

Purpose

This lesson involves the students brainstorming for ideas for good projects by considering the world around them. The students then have the opportunity to develop a topic of their choice by generating questions.

Time Allocation

One 40-minute period

Apparatus and Materials

None required

Suggested Approach

1. Write a few headings such as the kitchen, the garden, TV, toys, or the environment on the blackboard. Set the stage for brainstorming using the newspaper article. In brainstorming, no idea is a bad idea. The best ideas are selected afterward.

2. Invite the students to make a personal selection of an idea.

3. The sections on "Making Sauerkraut" and "Developing an Idea" offer two ways the students might develop the idea they have selected.

12*** IDEAS FROM THE SCIENTIFIC LITERATURE

Purpose

Lessons 9***, 10**, and 10*** have progressively introduced students to the scientific literature and the scientific community it serves. With these lessons as background, the literature becomes a natural starting point for ideas.

Time Allocation

One or two 40-minute periods

Apparatus and Materials

- Plastic bottle filled with a sand-and-water mixture

- Cornstarch-and-water mixture

Suggested Approach

1. It's probably sufficient to demonstrate the two phenomena with the extraordinary liquids.

2. Follow the demonstration with a discussion of how the students would investigate the phenomena.

3. With a view to encouraging students to carry out a search for good ideas (a homework assignment, perhaps), read the section that describes scientific journals.

References

Carlisle, R. W. 1989. Children as experimenters, SSHRC Award 410-88-0832, Faculty of Education, University of British Columbia.

Council of Ministers of Education, Canada (CMEC). 1997. *Common framework of science learning outcomes.* Toronto: Council of Ministers of Education, Canada. *http://publications.cmec.ca/ science/framework.*

Haysom, J., and M. Bowen. 2010. *Predict, observe, explain: Activities enhancing scientific understanding.* Arlington, VA: NSTA Press.

Medawar, P. B. 1969. Science and literature. *Encounter* (January): 15-23.

National Research Council (NRC). 1996. *National science education standards.* Washington, DC: National Academies Press.

National Research Council (NRC). 2012. *A framework for K–12 science education: Practices, crosscutting concepts, and core ideas.* Washington, DC: National Academies Press.

O'Connor, J. 1990. *SI² Network Newsletter, Faculty of Education, University of British Columbia* 2 (7).

Appendix A

*Blackline Master for Starting Point 1 and Lesson 3**

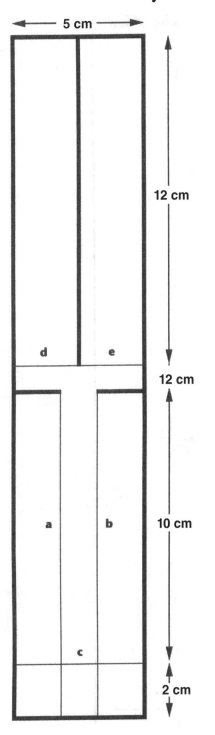

Instructions

1. Cut along thick solid lines.

2. Fold along thick lines a, b, and then c.

3. Attach paper clip to hold c in place.

4. Fold thin lines d and e in opposite directions.

Note: When photocopying, enlarge square 140% to get the proper dimensions on the copy.

Appendix B

Learning to Play the Numbers Game—Worksheet

Record your findings by circling your answer.

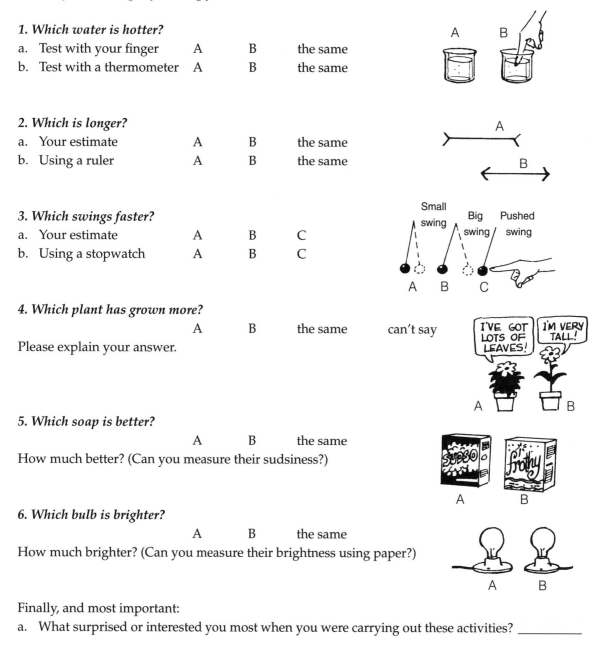

1. Which water is hotter?

a. Test with your finger   A   B   the same

b. Test with a thermometer   A   B   the same

2. Which is longer?

a. Your estimate   A   B   the same

b. Using a ruler   A   B   the same

3. Which swings faster?

a. Your estimate   A   B   C

b. Using a stopwatch   A   B   C

4. Which plant has grown more?

          A   B   the same   can't say

Please explain your answer.

5. Which soap is better?

          A   B   the same

How much better? (Can you measure their sudsiness?)

6. Which bulb is brighter?

          A   B   the same

How much brighter? (Can you measure their brightness using paper?)

Finally, and most important:

a. What surprised or interested you most when you were carrying out these activities? _____

_____

b. Which of the questions above do you find the most difficult to answer? Can you explain why?

_____

Index

Index

Index

Index